ENERGY CONSERVATION IN RESIDENTIAL, COMMERCIAL, AND INDUSTRIAL FACILITIES

ENERGY CONSERVATION IN RESIDENTIAL, COMMERCIAL, AND INDUSTRIAL FACILITIES

Edited by
HOSSAM A. GABBAR

Published by John Wiley & Sons, Inc., Hoboken, New Jersey.
Published simultaneously in Canada.

For general information on our other products and services or for technical support, please contact our Customer Care Department within the United States at (800) 762-2974, outside the United States at (317) 572-3993 or fax (317) 572-4002.

Wiley also publishes its books in a variety of electronic formats. Some content that appears in print may not be available in electronic formats. For more information about Wiley products, visit our web site at www.wiley.com.

Library of Congress Cataloging-in-Publication Data is available.

ISBN: 978-1-119-42206-8

Cover design: Wiley
Cover image: © WangAnQi/iStockphoto

Printed in the United States of America.

V10001888_062618

I dedicate this book to my wife Naila Gaber for her great support, and to my son John Gaber and daughter Sophia Gaber for inspiring and motivating me to complete the book.

TABLE OF CONTENTS

PART II ENERGY SYSTEMS

PART III   ENERGY CONSERVATION STRATEGIES

6   INTEGRATED PLANNING AND OPERATIONAL CONTROL OF RESILIENT MEG FOR OPTIMAL DERs SIZING AND ENHANCED DYNAMIC PERFORMANCE   205

Hossam A. Gabbar, Ahmed M. Othman, and Aboelsood Zidan

7   PERSPECTIVES OF DEMAND-SIDE MANAGEMENT UNDER SMART GRID CONCEPT   225

Onur Elma and Hossam A. Gabbar

PREFACE

Energy consumption in infrastructures represents almost one-third of total energy demand. As energy is linked to greenhouse gas emissions, which is linked to climate change and global warming, it is important to provide intelligent systems to support both energy conservation and energy supply in infrastructure systems.

This book shows business model and engineering design framework for practical implementation of energy conservation in infrastructures such as buildings, hotels, public facilities, industrial facilities, transportation, and water/energy supply infrastructures. Key performance indicators are modeled and used to evaluate energy conservation strategies and energy supply scenarios as part of the design and operation of energy systems in infrastructures. The proposed system approach shows effective management of building energy knowledge, which supports the simulation, evaluation, and optimization of several building energy conservation scenarios. Case studies are used to illustrate the proposed energy conservation framework, practices, methods, engineering designs, control, and technologies.

This book will offer the following new concepts:

- Infrastructure energy modeling
- Building envelope modeling
- Energy conservations methods
- Energy semantic networks (ESN) superstructures
- Energy conservation strategies and performance measures
- Examples in HVAC, lighting, appliances, storage, and machines

- Energy conservation optimization techniques
- Risk-based life cycle assessment
- Control strategies and systems for energy conservation
- Advanced energy audit systems

This book is structured into four parts:

Part I Energy Infrastructure Systems
Part II Energy Systems
Part III Energy Conservation Strategies
Part IV Resiliency, Protection, Control, and Optimization Systems

This book will help technology providers, infrastructure support industries, construction companies, municipalities, and regulatory institutions to study and manage energy conservation in infrastructures that include residential buildings, industrial facilities, transportation, and city infrastructures.

University of Ontario Institute of Tech, Ontario, Canada   Hossam A. Gabbar

AUTHORS' BIOGRAPHY

Onur Elma received his M.S. and Ph.D. degrees in Electrical Engineering from Yildiz Technical University (YTU), Istanbul, Turkey, in 2011 and 2016, respectively. He worked as a project engineer in the industry between 2009 and 2011. He has been employed as Research Assistant in Electrical Engineering Department in YTU since 2011. He has been in Smart Energy Research Center (SMERC) at University of California, Los Angeles (UCLA) as a visiting researcher from 2014 to 2015. Currently, he is working as a post-doc researcher at University of Ontario Institute of Technology (UOIT). He has participated in many national and international projects and also has to his credit more than 25 papers. His research interests include smart grid, electric vehicles, home energy management systems, and renewable energy systems.

Hossam A. Gabbar is a full Professor in Faculty of Energy Systems and Nuclear Science at University of Ontario Institute of Technology (UOIT), and cross appointed in the Faculty of Engineering and Applied Science, where he has established both the Energy Safety and Control Lab (ESCL) and Advanced Plasma Engineering Lab. He is the recipient of the Senior Research Excellence Award for 2016, UOIT. Dr. Gabbar obtained his B.Sc. (Honors) degree in 1988 in first class from the Faculty of Engineering, Alexandria University (Egypt). In 2001, he obtained his Ph.D. degree from Okayama University (Japan) in Safety Engineering. From 2001 to 2004, he worked at Tokyo Institute of Technology (Japan), and from 2004 to 2008, he was Associate Professor in the Division of Industrial Innovation Sciences at Okayama University (Japan). From 2007 to 2008, he was a Visiting

Professor at the University of Toronto. He has more than 210 publications to his credit, including patents, books/chapters, journals, and conference papers.

Shibo Luo was born in Hunan, China, in 1977. He is currently pursuing the Ph.D. degree in Shanghai Jiao Tong University, Shanghai, China. He participates in many national projects, such as National Natural Science Foundation of China, National "973" Planning of the Ministry of Science and Technology, China, and so on. His research interests include SDN network security, network service composition, and so on.

Farayi Musharavati is currently Associate Professor in Department of Mechanical and Industrial Engineering, Qatar University. He obtained his Ph.D. in Manufacturing Systems from University Putra Malaysia in 2008. He holds MSc degree in both Manufacturing Systems and Renewable Energy and a B.Tech. (Honors) degree in Mechanical and Production Engineering from the University of Zimbabwe, Zimbabwe. Research interests include manufacturing systems, energy management, sustainability, waste management, life cycle assessment, applications of computational intelligence, smart water and smart energy, and renewable energy applications.

Ahmed M. Othman is Associate professor at Zagazig University, Egypt. He worked as a postdoctorate fellow at University of Ontario Institute of Technology (UOIT), Canada. He obtained his B.Sc. and M.Sc. degrees in electrical engineering from Zagazig University, Egypt in 2002 and 2004, respectively, and received his Ph.D. degree in electrical engineering from Aalto University, Finland in 2011. His current research areas include power quality issues, DFACTS technology, distributed energy resources interface and control and application of artificial intelligent techniques on power systems, microgrid, and renewable energy.

Shaligram Pokharel is a professor of Mechanical and Industrial Engineering at Qatar University, Doha, Qatar. Prior to joining this university, he held academic positions in Nanyang Technological University, Singapore. He holds B.E. (Honors) in Mechanical Engineering from the Regional Engineering College (Kashmir, India) and M.A.Sc. and Ph.D. in Systems Design Engineering from the University of Waterloo, Ontario, Canada. His research areas are focused in energy planning and modeling, low carbon supply chains, engineering management, reverse logistics, and emergency and humanitarian logistics.

Jason Runge obtained his M.A.Sc degree in Electrical Engineering in 2016 and a Bachelor of Engineering in Energy Systems Engineering in 2014 from the University of Ontario Institute of Technology. Currently, he is working toward his Ph.D. in Building Engineering at Concordia University. His research interests include energy forecasting, energy management, building management systems, and renewable energy systems.

Khairy Sayed received his B.S. degree in Electrical Power and Machines in 1997 from Assiut University, Assiut, Egypt. He obtained his Master's degree from the Electrical Energy Saving Research Center, Graduate School of Electrical Engineering, Kyungnam University, Masan, Korea, in 2007. He received his Ph.D. degree from Assiut University in 2013. He is working as Assistant Professor in the department of Electrical Engineering, Sohag University, Egypt. His research interests include soft switching converters, solar PV, wind energy, fuel cell, power conditioners for renewable energy sources, smart energy grids, protection, and control of smart microgrids. He has more than 10 years of experience in SCADA/DMS during his work in Middle Egypt Electricity Distribution company as a system integrator for control center project. He was a Visiting Scholar in University of Ontario Institute of Technology (UOIT) in 2016. At present he is working as a head of electrical department in Assiut Integrated Technical Education Cluster (ITEC), Assiut, Egypt.

Kartikey Singh is a final year student in electrical engineering and visiting student at the Faculty of Energy Systems and Nuclear Science, University of Ontario Institute of Technology, where he worked on a research project in the area of transportation electrification and supported research tasks related to heuristic approaches for central control system design for resilient microgrids in transportation electrification applications.

Jun Wu received the Ph.D. degree in Information and Telecommunication Studies from Waseda University, Japan, in 2011. He was Post-Doctoral Researcher at the Research Institute for Secure Systems, National Institute of Advanced Industrial Science and Technology (AIST), Japan, from 2011 to 2012. He was Researcher at the Global Information and Telecommunication Institute, Waseda University, Japan, from 2011 to 2013. He is currently Associate Professor of the School of Electronic Information and Electrical Engineering, Shanghai Jiao Tong University, China. He is also the Vice Director of National Engineering Laboratory for Information Content Analysis Technology, Shanghai Jiao Tong University, China. He is the chair of IEEE P21451-1-5 Standard Working Group. His research interests include advanced computing, communications and security techniques of software-defined networks (SDN), information-centric networks (ICN) smart grids, and Internet of Things (IoT) on which he has published more than 90 refereed papers. He has been Guest Editor of the *IEEE Sensors Journal*. He is Associate Editor of the IEEE Access. He is a member of IEEE.

Aboelsood Zidan was born in Sohag, Egypt, in 1982. He received his B.Sc. and M.Sc. degrees in electrical engineering from Assiut University, Egypt, in 2004 and 2007, respectively, and his Ph.D. in electrical engineering from University of Waterloo, Waterloo, Ontario, Canada in 2013. He is currently Assistant Professor at Assiut University, Egypt. His research interests include distribution automation, renewable energy, distribution system planning, and smart grids.

LIST OF CONTRIBUTORS

ONUR ELMA, Department of Electrical Engineering, Yildiz Technical University, Istanbul, Turkey; Faculty of Energy Systems and Nuclear Science, University of Ontario Institute of Technology, Oshawa, Canada

HOSSAM A. GABBAR, Faculty of Energy Systems and Nuclear Science, University of Ontario Institute of Technology; Faculty of Engineering and Applied Science, University of Ontario Institute of Technology, Oshawa, Canada

SHIBO LUO, School of Electronic Information and Electrical Engineering, Shanghai Jiao Tong University, Shanghai, China

FARAYI MUSHARAVATI, Department of Mechanical and Industrial Engineering, College of Engineering, Qatar University, Doha, Qatar

AHMED M. OTHMAN, Faculty of Energy Systems and Nuclear Science, University of Ontario Institute of Technology, Oshawa, Canada; Electrical Power and Machines Department, Faculty of Engineering, Zagazig University, Zagazig, Egypt

SHALIGRAM POKHAREL, Department of Mechanical and Industrial Engineering, College of Engineering, Qatar University, Doha, Qatar

JASON RUNGE, Faculty of Engineering and Applied Science, University of Ontario Institute of Technology, Oshawa, Canada

KHAIRY SAYED, Sohag University, Egypt; Faculty of Energy Systems and Nuclear Science, University of Ontario Institute of Technology, Oshawa, Canada

KARTIKEY SINGH, Faculty of Energy Systems and Nuclear Science, University of Ontario Institute of Technology, Oshawa, Canada

JUN WU, School of Electronic Information and Electrical Engineering, Shanghai Jiao Tong University, Shanghai, China

ABOELSOOD ZIDAN, Faculty of Energy Systems and Nuclear Science, University of Ontario Institute of Technology, Oshawa, Canada

ACKNOWLEDGMENTS

The editor would like to thank all contributors to this book, and the research team at the Smart Energy Systems Lab (SESL) at UOIT for their full dedication and quality research. Also, the editor would like to thank IEEE SMC for providing the chance to publish this work. We acknowledge UOIT for their continuous support to the research work at SESL.

PART I

ENERGY INFRASTRUCTURE SYSTEMS

CHAPTER 1

ENERGY IN INFRASTRUCTURES

HOSSAM A. GABBAR[1,2]
[1]Faculty of Energy Systems and Nuclear Science, University of Ontario Institute of Technology, Oshawa, Canada
[2]Faculty of Engineering and Applied Science, University of Ontario Institute of Technology, Oshawa, Canada

1.1 INFRASTRUCTURE SYSTEMS

As measured in 2015, around 1.2 billion people, constituting 17% of the global population, do not have electricity, and 2.7 billion people, constituting 38% of the global population, have risks on their health due to the reliance on the traditional use of biomass for cooking [1].

In order to discuss energy systems and conservation strategies in infrastructures, it is essential to analyze the infrastructure physical systems and their types, classifications, and energy requirements. It is possible to find a suitable definition of infrastructures as the fundamental facilities and systems that serve a region, area, community, city, or country, including the support facilities such as utilities, services, and transportation that are necessary for the economic development and perform all necessary functions. There are number of ways to classify infrastructures, such as size, criticality, use, occupancy, location, and surroundings. Infrastructures can support residential functions, commercial and public functions, transportation functions (including land, sea, air), and industrial functions. Infrastructures can be viewed as system of systems; for example, infrastructures include communications and cyber security, computational/technological, waste management, emergency and disaster management, defense and military, and other supporting infrastructures. The better we understand infrastructures, the better we design and operate energy systems in these infrastructures. Infrastructure modeling should support design and operational activities, with appropriate and comprehensive performance measures to evaluate design and operation features

Energy Conservation in Residential, Commercial, and Industrial Facilities, First Edition.
Edited by Hossam A. Gabbar.
© 2018 The Institute of Electrical and Electronics Engineers, Inc. Published 2018 by John Wiley & Sons, Inc.

and alternatives. Requirement analysis of infrastructures should include energy demand, risk management, performance, and sustainability requirements.

1.1.1 Infrastructure Classifications

Energy use in infrastructures can be controlled and optimized based on the nature of loads and energy systems implemented in these infrastructures. For proper planning, design, and operation of energy systems to support these infrastructures, it is important to analyze the classifications of infrastructures. Figure 1.1 shows hierarchical classification of infrastructures based on nature, type, use, function, and energy requirements. There are interrelations among these infrastructures, for example, water infrastructures are linked to residential, industrial, and commercial. Similarly, energy and waste are linked to all other infrastructures.

In order to understand energy consumption in different regions, power consumption in Ontario has been selected, as presented in Figure 1.2, where it shows the consumption in residential, commercial, industrial, electric vehicle, transit, and others. Power consumption in residential is very close to that consumed in commercial, while industrial is the third dominating sector for power consumption.

1.1.2 Infrastructure Systems

Infrastructure system includes technical and technological infrastructures to support all functions and the management of life cycle activities in infrastructures including flow and control of information across all elements of the infrastructure systems. Modeling of processes of infrastructure systems includes players, roles,

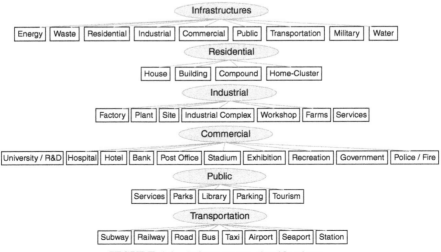

FIGURE 1.1 Infrastructure classifications.

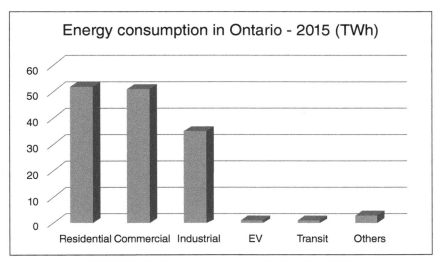

FIGURE 1.2   Power consumption in Ontario – 2015.

physical systems, functional modeling, financial modeling, planning, engineering design, operation, and management practices. One major component of infrastructure systems is the safety and protection systems to ensure the resiliency against hazardous, emergencies, and disaster situations and to sustain the stated target functions from the infrastructure systems.

1.2   ENERGY SYSTEMS IN RESIDENTIAL FACILITIES

Energy consumption in residential facilities constitutes one of the largest consumption of energy in cities and communities in Canada and worldwide. In 2015, energy consumption in residential facilities in Ontario is 52 TWh, which represents 36% of total energy consumption. Energy consumption in residential facilities include heating/cooling, electric loads, water heating, laundry, dishwashing, refrigerators and freezers, cooking, TV, lighting, and computer-related equipment, as shown in Figure 1.3. The highest energy use is in heating and cooling and ventilation, where it is clear the reduced use from 2013 to 2040. This can be justified by improved heating and cooling technologies and efficiencies. Electric loads and water heating are second largest energy use in the residential sector. Energy conservation strategies are widely adopted by utilities to reduce energy demand from utilities in residential facilities. Typically, utility grids supply energy to residential facilities. Energy conservation can represent around 1–3% of total energy demand in residential facilities. With the penetration of local distributed generation, energy can be supplied by

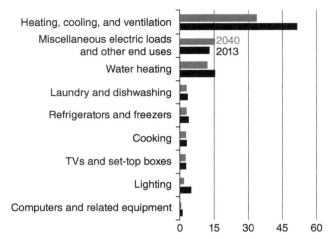

FIGURE 1.3   Residential sector delivered energy intensity for selected end uses in the Reference case, 2013 and 2040 (million Btu per household per year) [2].

renewable energy technologies such as PV, energy storage, wind, gas generators, fuel cells, and geothermal systems.

There are a number of energy systems and technologies that are adopted in residential facilities, such as gas-fired water heaters, oil-fired water heaters, electric water heaters, heat pump water heaters, instantaneous water heaters, solar water heaters, gas-fired furnaces, oil-fired furnaces, gas-fired boilers, oil-fired boilers, room air conditioners, central air conditioners, air-source heat pumps, ground-source heat pumps, gas-source heat pumps, electric resistance furnaces, electric resistance unit heaters, cordwood stoves, wood pellet stoves, refrigerators-freezers, freezers, natural gas cooktops and stoves, clothes washers, clothes dryers, and dishwashers. Among the factors that are used to evaluate these energy systems are capacity, efficiency, energy factor (EF), combined energy factor (CEF), annual energy use, annual water use, average life, retail equipment costs, installation costs, and maintenance costs. These factors are used to evaluate the different energy systems in residential facilities to ensure most effective technology that can be applied in different regions and weather conditions.

Energy consumption in residential facilities can be viewed as in Figure 1.4, where it shows different types of energy sources, such as propane, kerosene, distillate fuel oil, natural gas, renewable energy, and electricity.

It is clear that electricity and natural gas represent the highest consumption from 2012 and projected till 2040. It is also noted that losses are quite high and energy conservation strategies will be essential for effective savings.

Energy prices for residential use are shown in Figure 1.5, which shows price of natural gas (NG) is the lowest, while electricity price is the highest.

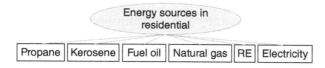

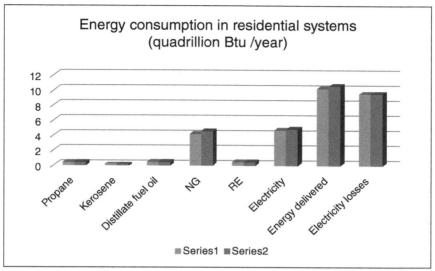

FIGURE 1.4   Energy consumption in residential systems, quadrillion Btu per year in the United States, 2012: gray, 2020: dark gray [2].

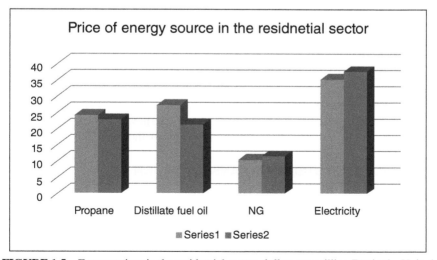

FIGURE 1.5   Energy prices in the residential sector, dollars per million Btu in the United States, 2012: gray, 2020: dark gray [2].

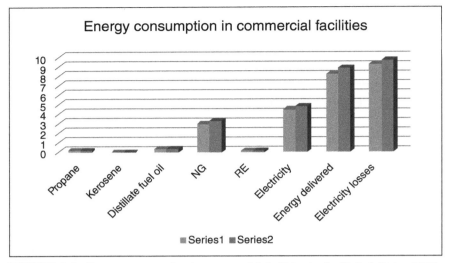

FIGURE 1.6 Energy consumption in commercial facilities in the United States, 2012: gray, 2020: dark gray [2].

1.3 ENERGY SYSTEMS IN COMMERCIAL FACILITIES

Energy consumption in commercial facilities, as stated by Department of Energy (DOE) [2], is shown in Figure 1.6. Electricity consumption is higher than NG use. While NG is cheaper than electricity, it is possible to provide better solution with increase in NG penetration in commercial use.

Also, energy prices in commercial facilities are shown in Figure 1.7.

It is shown that NG price for the residential sector is higher than NG price for the commercial sector.

1.4 ENERGY SYSTEMS IN INDUSTRIAL FACILITIES

In the industrial sector, Figure 1.8 shows the consumption from 2015 [2].

Also, energy prices in the industrial sector are shown in Figure 1.9, which shows the NG as the lowest clean energy source for the industrial sector.

1.5 ENERGY SYSTEMS IN TRANSPORTATION INFRASTRUCTURES

It is widely known that greenhouse gas (GHG) emission from transportation sector is high. The proper analysis of energy consumption in the transportation is important to address issues related to energy conservation with sustainability considerations, as shown in Figure 1.10.

Energy prices in the transportation sector are shown in Figure 1.11.

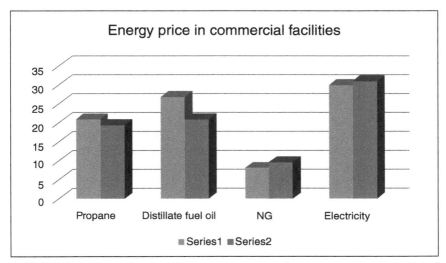

FIGURE 1.7   Energy prices in the commercial sector, dollars per million Btu in the United States, 2012: gray, 2020: dark gray [2].

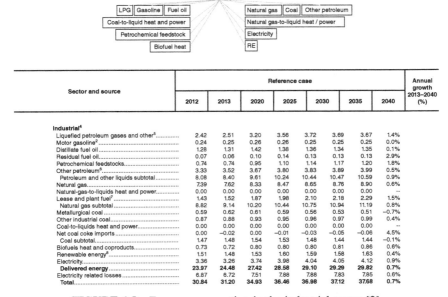

Sector and source	Reference case							Annual growth 2013–2040 (%)
	2012	2013	2020	2025	2030	2035	2040	
Industrial[4]								
Liquefied petroleum gases and other[5]	2.42	2.51	3.20	3.56	3.72	3.69	3.67	1.4%
Motor gasoline[9]	0.24	0.25	0.26	0.26	0.25	0.25	0.25	0.0%
Distillate fuel oil	1.28	1.31	1.42	1.38	1.36	1.34	1.35	0.1%
Residual fuel oil	0.07	0.06	0.10	0.14	0.13	0.13	0.13	2.9%
Petrochemical feedstocks	0.74	0.74	0.95	1.10	1.14	1.17	1.20	1.8%
Other petroleum[6]	3.33	3.52	3.67	3.80	3.83	3.89	3.99	0.5%
Petroleum and other liquids subtotal	8.08	8.40	9.61	10.24	10.44	10.47	10.59	0.9%
Natural gas	7.39	7.62	8.33	8.47	8.65	8.76	8.90	0.6%
Natural-gas-to-liquids heat and power	0.00	0.00	0.00	0.00	0.00	0.00	0.00	--
Lease and plant fuel[7]	1.43	1.52	1.87	1.98	2.10	2.18	2.29	1.5%
Natural gas subtotal	8.82	9.14	10.20	10.44	10.75	10.94	11.19	0.8%
Metallurgical coal	0.59	0.62	0.61	0.59	0.56	0.53	0.51	–0.7%
Other industrial coal	0.87	0.88	0.93	0.95	0.96	0.97	0.99	0.4%
Coal-to-liquids heat and power	0.00	0.00	0.00	0.00	0.00	0.00	0.00	--
Net coal coke imports	0.00	–0.02	0.00	–0.01	–0.03	–0.05	–0.06	4.5%
Coal subtotal	1.47	1.48	1.54	1.53	1.48	1.44	1.44	–0.1%
Biofuels heat and coproducts	0.73	0.72	0.80	0.80	0.80	0.81	0.86	0.6%
Renewable energy[8]	1.51	1.48	1.53	1.60	1.59	1.58	1.63	0.4%
Electricity	3.36	3.26	3.74	3.98	4.04	4.05	4.12	0.9%
Delivered energy	23.97	24.48	27.42	28.58	29.10	29.29	29.82	0.7%
Electricity related losses	6.87	6.72	7.51	7.88	7.88	7.83	7.85	0.6%
Total	30.84	31.20	34.93	36.46	36.98	37.12	37.68	0.7%

FIGURE 1.8   Energy consumption in the industrial sector [2].

Sector and source	Reference case							Annual growth 2013–2040 (%)
	2012	2013	2020	2025	2030	2035	2040	
Industrial[1]								
Propane	21.3	20.3	19.6	20.5	21.5	22.9	24.5	0.7%
Distillate fuel oil	27.4	27.3	21.2	23.5	26.1	29.2	32.7	0.7%
Residual fuel oil	20.6	20.0	13.3	15.1	17.2	19.7	23.5	0.6%
Natural gas[2]	3.8	4.6	6.2	6.9	6.8	7.5	8.8	2.5%
Metallurgical coal	7.3	5.5	5.8	6.2	6.7	6.9	7.2	1.0%
Other industrial coal	3.3	3.2	3.3	3.5	3.6	3.7	3.9	0.7%
Coal to liquids	--	--	--	--	--	--	--	--
Electricity	19.8	20.2	21.3	22.4	22.6	23.3	24.7	0.7%

FIGURE 1.9   Energy prices in the industrial sector, dollars per million Btu [2].

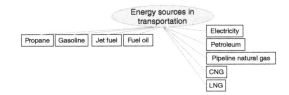

Sector and source	Reference case							Annual growth 2013–2040 (%)
	2012	2013	2020	2025	2030	2035	2040	
Transportation								
Propane	0.05	0.05	0.04	0.05	0.05	0.06	0.07	1.3%
Motor gasoline[2]	15.82	15.94	15.35	14.22	13.30	12.82	12.55	−0.9%
of which: E85[9]	0.01	0.02	0.03	0.12	0.20	0.24	0.28	10.0%
Jet fuel[10]	2.86	2.80	3.01	3.20	3.40	3.54	3.64	1.0%
Distillate fuel oil[11]	5.80	6.50	7.35	7.59	7.76	7.94	7.97	0.8%
Residual fuel oil	0.67	0.57	0.35	0.36	0.36	0.36	0.36	−1.6%
Other petroleum[12]	0.15	0.15	0.16	0.16	0.16	0.16	0.16	0.2%
Petroleum and other liquids subtotal	25.35	26.00	26.27	25.57	25.03	24.88	24.76	−0.2%
Pipeline fuel natural gas	0.75	0.88	0.85	0.90	0.94	0.94	0.96	0.3%
Compressed/liquefied natural gas	0.04	0.05	0.07	0.10	0.17	0.31	0.71	10.3%
Liquid hydrogen	0.00	0.00	0.00	0.00	0.00	0.00	0.00	--
Electricity	0.02	0.02	0.03	0.04	0.04	0.05	0.06	3.4%
Delivered energy	26.16	26.96	27.22	26.60	26.18	26.19	26.49	−0.1%
Electricity related losses	0.05	0.05	0.06	0.07	0.08	0.10	0.12	3.1%
Total	26.20	27.01	27.29	26.67	26.27	26.29	26.61	−0.1%

FIGURE 1.10   Energy consumption in the transportation sector [2].

Sector and source	Reference case							Annual growth 2013–2040 (%)
	2012	2013	2020	2025	2030	2035	2040	
Transportation								
Propane	25.3	24.6	24.0	24.7	25.5	26.5	27.6	0.4%
E85[3]	35.7	33.1	30.4	29.0	31.2	33.2	35.4	0.3%
Motor gasoline[4]	30.7	29.3	22.5	24.3	26.4	29.1	32.3	0.4%
Jet fuel[5]	23.0	21.8	16.1	18.3	21.3	24.5	28.3	1.0%
Diesel fuel (distillate fuel oil)[6]	28.8	28.2	23.1	25.5	28.0	31.1	34.7	0.8%
Residual fuel oil	20.0	19.3	11.7	13.3	15.4	17.6	20.3	0.2%
Natural gas[7]	20.4	17.6	17.8	16.8	15.7	17.1	19.6	0.4%
Electricity	27.8	28.5	30.2	32.3	32.9	33.9	36.0	0.9%

FIGURE 1.11   Energy prices in the transportation sector, dollars per million Btu [2].

1.6   ENERGY PRODUCTION AND SUPPLY INFRASTRUCTURES

Energy demand and the associated loads include power/electricity, thermal, fuel, water, and their links to the required work. Due to the variations in these loads, energy production and supply chains should provide adequate flexibility and adaptation to local and regional energy needs.

Energy production and supply chains are integrated with R&D chains to support development and implementation of advanced energy systems in different infrastructures, as shown in Figure 1.12.

Energy life cycle includes energy sources development and treatment, energy conversion, storage and transportation, and energy utilization. This includes all energy sources, fossil fuels, renewables, emerging energy technologies, hydrogen, bioenergy, and nuclear energy. The development of energy systems for different infrastructures might vary, based on resources availability, nature of infrastructures, weather, external and geopolitical factors, and the regional requirements. The proposed analysis requires accurate estimation of energy demand, consumption, and load profiles, which are the basis for proper energy system development.

The presented energy production and supply infrastructures show different sources and technologies ranging from hydrogen, biomass/biofuel, nuclear, thermal power, hydropower, wind, photovoltaic, geothermal, and other emerging energy technologies. All are mapped via energy infrastructures to supply energy to different loads such as residential, commercial, and industrial facilities, as well as transportation networks.

With the advancement in energy conversion and generation technologies, it is possible to cover energy needs with a combination of electricity, thermal, fuel, and even water. For example, to cover heating requirements, we can have CHP to provide heat, or electric heater, or heat pumps with water circulation. The selection of the type of energy technology and required amount of energy supply will vary,

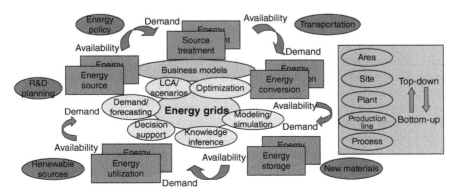

FIGURE 1.12   Life cycle engineering of energy in infrastructures.

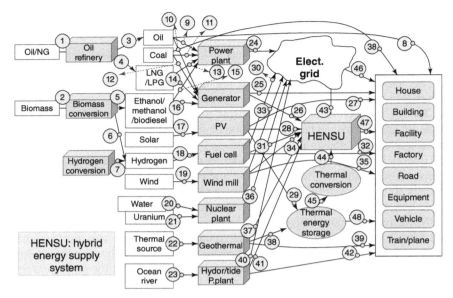

FIGURE 1.13   Energy production and supply infrastructures.

based on number of factors such as installation/operation costs, source/supply availability and costs, infrastructure requirements/limitations, and other corporate and regional requirements and regulations. Figure 1.13 shows possible energy infrastructures from different sources to variety of loads with different sizes and scales. It is possible to plan energy supply in infrastructures based on different scenarios such as electricity grids, natural gas grids, hydrogen grids, and/or mixture of these sources along with emerging energy technologies.

1.7   CONCLUSION

This chapter presented a summary of energy sources and their deployment in infrastructures, where energy is the backbone of improved infrastructure performance. This includes residential, commercial, industrial, and transportation and energy infrastructures. This chapter presented analysis and study of energy in infrastructures with all infrastructure classifications. The corresponding energy expenditures for each sector is viewed in Figure 1.14, which shows highest expenditures are in the transportation sector, followed by the residential sector. This introductory chapter explored energy in infrastructures and the corresponding classifications that will support energy conservation planning, operation, and control.

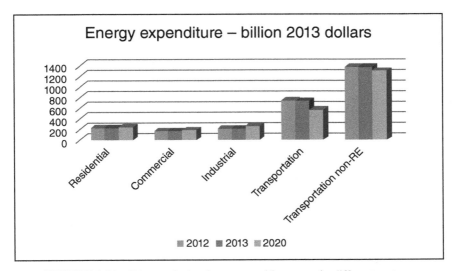

FIGURE 1.14   Price analysis of nonrenewable energy in different sectors.

REFERENCES

1. World Energy Outlook (2015) International Energy Agency.
2. Annual Energy Outlook (2015) DOE/EIA-0383(2015).

CHAPTER 2

BUILDING ENERGY MANAGEMENT SYSTEMS (BEMS)

KHAIRY SAYED[1,2] and HOSSAM A. GABBAR[2,3]
[1]Sohag University, Egypt
[2]Faculty of Energy Systems and Nuclear Science, University of Ontario Institute of Technology, Oshawa, Canada
[3]Faculty of Engineering and Applied Science, University of Ontario Institute of Technology, Oshawa, Canada

2.1 INTRODUCTION

Building energy management systems (BEMS) are computer-based control systems that control and monitor the mechanical and electrical equipment in buildings such as ventilation, heating, lighting, power systems, and so on. This is sometimes called building management systems (BMS); they connect the building services plant back to a central computer to enable control of on/off times, humidity, temperatures, and so on. Data cables connect the controlled plant through a series of hubs called outstations around the building back to a master station that is central supervisor computer where building operators can supervisory control and monitor the building. Energy management systems can save millions on annual energy bills while increasing prosperity in your building and making it easier to run. Today's energy management systems (EMS) make managing energy utilization (and bills) easier than ever.

Energy management systems allow your facility to power equipment only when needed. For many facilities, this eliminates the waste of lighting, heating, and cooling portions of the building that are not used around-the-clock. Optimized controls enhance your building's current mechanical systems and increase your ability to manage comfort and air quality throughout the building. By reducing unnecessary use of equipment, energy management systems can prolong the life of your building's mechanical and lighting systems, and reduce maintenance costs [1].

Energy Conservation in Residential, Commercial, and Industrial Facilities, First Edition.
Edited by Hossam A. Gabbar.
© 2018 The Institute of Electrical and Electronics Engineers, Inc. Published 2018 by John Wiley & Sons, Inc.

A wide variety of systems and methodologies have been proposed in the literature to address the issue of decreasing energy consumption in residential and commercial buildings [2–5]. These proposals are based on different yet complementary perspectives, and often take an interdisciplinary approach, which makes it hard to obtain a comprehensive view of the state of the art in the energy management of buildings.

The lack of a structured and unifying view over the available approaches and methodologies to be adopted during the design of such energy-aware systems was the main trigger for undertaking the research underlying this survey. We specifically focused on the underlying architectures and methodologies, as well as on the necessary techniques that go beyond the well-established *smart home* paradigm, thus progressing toward intelligent building management systems (BMSs), in accordance with the ambient intelligence (AMI) vision. The ideal application scenario for AMI considers the user as the focus of a pervasive environment augmented with sensors and actuators, where an intelligent system monitors environmental conditions and takes proper actions to satisfy user requirements [6]. AMI systems are characterized by a low intrusiveness and by the capability to adapt themselves to the users' behavior and to anticipate their requirements. In the specific context of a BMS for energy saving, this visionary goal becomes even more complex due to the presence of contrasting goals, that is, a satisfaction of user requirements by minimization of the consumed electrical energy. Throughout this review of literature, the main components constituting a BMS will be identified, namely, a sensory infrastructure for monitoring energy consumption and environmental features, a data processing software for processing sensory data and performing energy-saving strategies, a subsystem for user interactive interface, and an actuation infrastructure for modifying the environmental state. The different solutions presented in the literature will analyze for each component. Whenever possible, qualitative comparisons of various approaches will be provided with respect to their specific features.

To qualitatively evaluate different BEMS, a set of relevant characteristics will be identified. Through this assessment procedure, the end users have a relevant role; besides being affected by too tough energy-saving policies, users might be bothered by other structural features, such as a set of instrumental devices, or by algorithmic features, such as learning methods that force them to have a continuous interaction. However, we will refer to these topics such as the "user comfort," and we will underline the characteristics of various BEMS solutions in terms of scalability and complexity of the system architecture [7], the intrusiveness of the deployed sensory and actuating devices, and the expected influence of technology on user comfort.

The BEMS software provides monitoring, control functions, and alarms and enables the operators to optimize building performance. BEMS are vital components for managing energy demand, in particular in large complex buildings and multibuilding sites [8].

Digital and analog input signals transfer to the BEMS the values of temperature, humidity, and so on the building is running at. Inputs might also contain whether

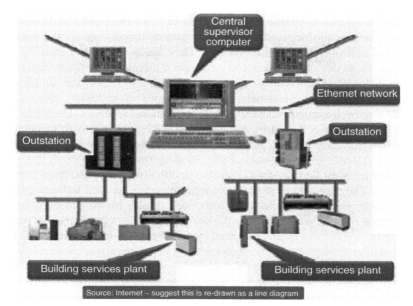

FIGURE 2.1   Typical structure of a BEMS.

equipment like fans, pumps, and boilers are working or not. Analog/digital outputs then transmit signals from the central supervisory controller PC to site equipment such as pumps, valves, fans, and so on to command their settings or to switch devices on and off, resulting in variation in comfort conditions. BEMS can be used to control very nearly anything and it is being widely used to control lighting and to monitor critical systems. Outstations programs automatically provide the local hubs to connect these inputs and outputs into the central supervisory master station (see Figure 2.1).

This enables the operator to program when things automatically switch on and off and what setting they operate at, for example, temperature, pressure, and humidity. A BEMS is really a tool for monitoring and controlling the building and a good operator can use the BEMS to optimize settings to maximize energy saving without compromising comfort and services. The outstations plants are usually linked through a local area network (LAN). Software normally support a graphical user interface GUI or human–machine interfaces HMI that is based on images of the plant being controlled. These dynamic displays offer real-time plant operation conditions that give an instantaneous window on what is happening in the building, which is called event loggers. As a core function, BEMS would control heating system, boilers, and pumps and then locally control the mixture of heat to achieve the optimum required room temperature. In air-conditioned buildings, BEMS would control chillers, cooling systems, and the systems that spread air throughout the building (for example, by operating fans or opening/closing dampers). BEMS can also control lighting or any other energy-consuming equipment and can also be used to log energy meters.

Modern systems have distributed intelligence in the outstations and also allow multisite data acquisition and control with remote monitoring via the wireless, telephone network, and satellite systems. They are increasingly becoming connected to smart devices like palmtop devices and mobile phones with alarms that tell on-call staff of problems or events in the building. BEMS can significantly promote the overall management and performance of buildings, promoting a holistic approach to controls and supporting operational feedback. Energy savings of 10–20% can be achieved by installing a BEMS compared with independent controllers for each system. However, BEMS cannot recompense for wrongly designed systems, incorrect maintenance, or poor management.

BEMS are ideal for providing control of multibuilding sites and large complex buildings. They are also used by large organizations to control buildings spread across wide areas like whole local authorities, health trusts, and even remote buildings across the whole country. Modern systems have intelligent outstations that can be investigated locally in a plant room to track down local problems. They can also have wireless connections to some devices to reduce or avoid cabling. A BEMS needs to be well specified and designed, with good documentation and an intuitive graphical user interface if it is to be used efficiently. In very small buildings, it is possible to achieve reasonable control using stand-alone controls for lighting, heating, and so on and this may be a cheaper option than a full BEMS. However, costs of controls have come down such that mini-BEMS are now competitive and hybrid systems that interconnect a series of local controllers are also available. So BEMS can be considered for controlling almost any size of buildings, but the improvement in management really becomes apparent in large distributed and complex sites/buildings. Ensuring good user interfaces with a BEMS is essential. Modern BEMS can be accessed in a number of ways (see Figure 2.2), for example, through web browsers via the Internet, through hand-held tablets and laptops, or through palm devices and smart mobile phones. Providing convenient access routes allows building operators to use the BEMS in a way that fits their role and the way they work and encourages them to utilize the system as a building optimization tool. Poor access or a lack of feedback normally results in facilities managers leaving the BEMS ignored in a corner of the operations room as a silent controller rather than a window into the building's performance. To optimize internal conditions and make ongoing savings, BEMS need to be regularly maintained. BEMS settings need to be checked at least every month and check that settings meet actual building requirements. When inspecting the system, focus on the following:

- *General*: Check the integrity of any cabling and connections and any cabinets or panels in the installation.
- *Sensors*: Test accuracy and review the suitability of their locations.
- *Actuators*: Examine control outputs and ensure that controlled devices are working over their full operating range.
- *Digital inputs*: Confirm that inputs are operational and working correctly.

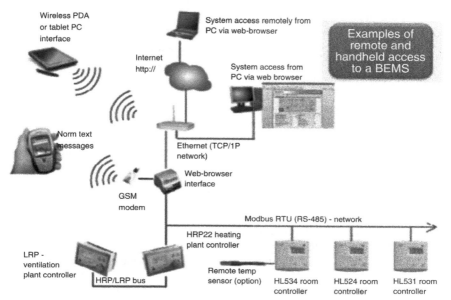

FIGURE 2.2   Typical BEMS user interfaces.

Calibrate or adjust switching devices if necessary:

- *Controllers:* Verify that battery supplies are adequate and that controllers automatically restart the following interruption to power supplies.
- *Record-Keeping:* Document key changes to the BEMS, including any alterations to set points and control strategies, software upgrades, additions to the network, any faults identified, or maintenance performed.

Maintaining controls really matters and underpins the building performance. The preferred maintenance regimes need to be determined at the beginning of the project. The important question to be answered by the client is whether they want

- an independent installation (independent of the control manufacturer) with a separate maintenance contract that can be moved according to contractual performance.

It is important that the BEMS maintenance contractor is consulted at the start of the build, during the design, and retained by the client to provide maintenance for the finished building. The on-site maintenance team can then have a good relationship with the controls subcontractor, allowing them to use continuous commissioning and rectify faults quickly.

BEMS can achieve lower running costs, improved comfort, maintenance, and building management through better system feedback on the performance of building on energy utilization and comfort. The monitoring facilities of a BEMS enable monitoring of plant status, environmental conditions, and energy utilization, providing the building operator with a real-time reporting of the building operation process. This can often lead to the identification of problems that may have gone unnoticed, for example, high-energy usage or plant left running continuously. Energy meters connected to a BEMS, providing real-time energy consumption patterns and ultimately a historical record of the buildings energy performance, can be logged and analyzed in a number of ways, both numerically and graphically. BEMS can, therefore, improve management information by trend logging performance, benefiting forward planning/costing. This can also encourage greater awareness of energy efficiency among staff. Energy efficiency improvements of 10–20% are common. However, it is important to establish the suitability of existing buildings and equipment to ensure the maximum savings. For a BEMS to work effectively in an existing building, it must be possible to zone the heating, ventilation, and lighting systems according to the use made in different areas.

The main advantage of a BEMS installation is the ease with which users can review the performance of controls and conveniently make adjustments. Other advantages include the following:

- Close control of environmental conditions, providing better comfort for occupants.
- Energy-saving control functions that will reduce energy bills (e.g., weather compensation).
- Ability to log and archive data for energy management purposes.
- Provision of events or rapid information on plant status.
- Automatic generation of system alarms for equipment failure or violation of normal condition.
- Identification of both planned and unplanned maintenance requirements (e.g., systems can record the number of hours that motors have run, or identify filters on air supply systems that have become blocked).
- Ease of expansion to control other plant, spaces, or buildings.

Once a BEMS has been installed and fully commissioned properly, it can be used as a tool to optimize building performance. Even the best designed and commissioned control strategy is likely to evolve with the user's and the building's requirements. A well-trained BEMS operator can carry out regular reviews of BEMS settings to gradually reduce room set points, operating times, and energy consumption without compromising comfort conditions.

This fine-tuning of the building controls often requires one or two full heating seasons to reach optimum settings. But the process does not end here: As the building usage and requirements change, so will set points and times; so this optimization is a continuous process as the building use changes.

This optimization process is particularly important where BEMS are controlling large multibuilding sites and buildings spread across a wide area. The BEMS operator can keep a watchful eye on operations and energy use from afar without having to visit the buildings. This central BEMS bureau approach is highly cost-effective and common in large estates and through FM providers.

As a result of this continuous optimization, it is important to maintain records of all changes to the system during the lifetime of the building with good reasons as to why changes have been made. Too many buildings have high operating hours and set points that have been badly programmed many years ago often as a result of occupant complaints. It is still very common to find buildings fully *on* running everything at high levels for 24 × 7, where just a little optimization can save a lot of energy, money, and carbon emissions with little or no investment.

A BEMS is only as good as the people who use it. It is essential that staff who will be operating and maintaining the system are trained appropriately. All reputable BEMS suppliers can provide and do encourage training as it is in their interest that the system works well. If installing a new BEMS involves key staff at the beginning of the project, ensure that they are aware of what the system can do and how to keep it performing efficiently.

Access through mobiles: It is essential to train staff to use the BEMS as a tool to manage the building. Ensuring staff have easy access through mobile devices can encourage this. The greater the understanding, the more likely the energy savings. This will involve training on the BEMS hardware and the software built into the BEMS. BEMS is a powerful tool for managing buildings, but it is still only as good as the staff operating it! All staff with access to the BEMS should develop experience in managing the building using it on a routine basis. Most BEMS have alarms set and staff should know what to do when these alarms show on the central supervisor.

As discussed earlier, in existing systems an annual review of control settings is essential and also important to ensure that the system is optimized in relation to the occupancy and requirement of the building. However, too many building operators leave this to the maintenance contractor under an annual contract. This often results in the building management relinquishing their responsibilities to the BEMS contractor and the building gradually drifts away from optimized settings. The provision for future retraining in the event of staff changes is very important to minimize this day-to-day reliance on suppliers for simple maintenance measures. Ensuring suitable BEMS user documentation for system fault finding and maintenance also plays a key part in this common mistake.

A BEMS installation is very site specific. Larger systems may require a feasibility study to identify the size, shape, and complexity of the BEMS required. This will establish what is to be controlled and monitored, the connections, hardware, and cabling required and the resulting benefits. It will also establish the architecture of the system, ring-shaped, star-shaped, and so on, and the location and capacity required in the outstation. The financial justification for a BEMS should ideally include a full life cycle costing calculation based on discounted cash flow. Estimates of potential savings should, where possible, account for contributions from improved maintenance and increased reliability, in addition to reduced energy consumption.

Planning and designing good controls at the outset is essential to achieving a good building. A client's brief for a good control system aims for energy efficiency while maintaining comfort. Designers' specifications need to set out the key energy features, so contractors appreciate what the control system needs to do. Low carbon buildings are best achieved when clients state an aim to have a low carbon building in operation in the client brief. The design, selection, installation, and operation of the resultant control system relate directly to these initial statements. Without such clear directions to the design team, a low carbon building is seldom achieved.

The scope for system expansion at each outstation should be carefully considered. Often the addition of a single point may require a complete outstation at considerable cost if all points on the original are occupied. If you already have a BEMS, then an upgrade or even extending it may bring very significant advantages. Really old systems may well need full replacement and may no longer be supported by the manufacturers. It is possible to connect meters to a BEMS for logging energy to provide a valuable tool for identifying savings. However, where larger buildings/sites are being submetered, it may often be better to have a dedicated automatic meter reading system with specialist software for meter logging, analysis, and reporting. How well your BEMS performs is reliant on a clear brief, good design, followed by good installation/commissioning. Some BEMS manufacturers offer their own design/installation service and some may even insist on this; others work with approved contractors. Either way, you should ask for references from sites similar to your own.

2.2 BEMS (BMS) CONTROL SYSTEMS OVERVIEW

Building energy management systems are computer-based systems that help to manage, control, and monitor building technical services (HVAC, lighting, etc.) and the energy consumption by devices used in the building. They provide the information and the tools that building managers need both to understand the energy usage of their buildings and to control and improve their buildings' energy performance.

Building management system offers dependable and user-friendly building control solutions to commercial, education, health care, leisure buildings, and more as shown in Figure 2.3. This includes delivering the world's fully integrated building solution encompassing HVAC, lighting, and access products.

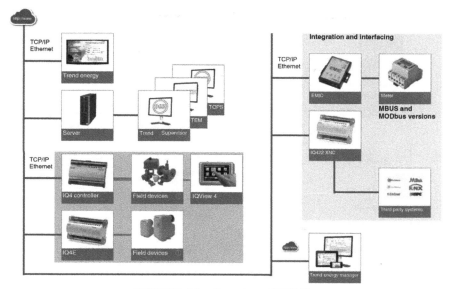

FIGURE 2.3   Overview of BEMS.

The building automation is geared toward energy management that has been one of the key concerns of every building owner. Having a healthy and productive environment while keeping your energy costs and your carbon footprint down makes your facility more valuable. Building automation offers complete end-to-end energy management solutions as shown in Figure 2.4. It helps in energy consumption reduction and save money.

The integrated building automation solutions are designed to simplify the management and protection of industrial, commercial, and residential buildings. These solutions answer to the rising emphasis on producing green facilities. BMS buildings can not only achieve certification but also perform at high levels with operational cost savings to property owners and managers. This solution can be implemented in both existing and new buildings. The building management system is an intelligent and integrated system for HVAC, fire, security, and access management in a building, as shown in Figure 2.5.

The BEMS sector is growing as organizations realize that it is one of the most effective solutions for optimizing energy efficiency in a building – providing the quick win that organizations are seeking. Energy managers and building owners might have previously installed systems in larger buildings, but new-generation technologies can cost-effectively extend the savings even to smaller buildings.

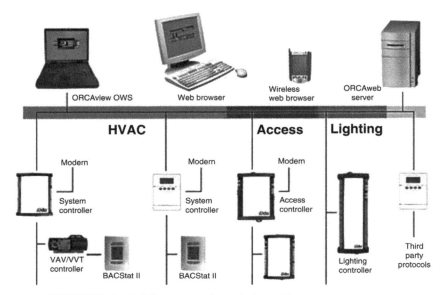

FIGURE 2.4 Building automation solution with energy management.

2.3 BENEFITS OF BUILDING ENERGY MANAGEMENT SYSTEMS

Greater energy efficiency and cost reduction is the main driver, but modern new systems also have the dual benefit of acting as an automatic monitoring and targeting system, monitoring, measuring, and analyzing consumption to assist with carbon reporting and managing efficiency. This enables users to collate, analyze, and transform these data into meaningful information, allowing them to monitor energy consumption, identify waste, and highlight areas for improvement and benchmark consumption against other similar buildings or organizations.

Automatic monitoring and targeting are particularly useful in monitoring multiple sites, enabling managers to gain both detailed analysis and a big picture overview of energy consumption across the business, while drilling down into specific locations. The reporting capabilities of some BEMS are necessary to meet increasing environmental legislation, such as the CRC energy efficiency scheme. The best systems will typically reduce consumption by 25% to deliver rapid payback on investment and reduce taxes on carbon emissions. Return on investment for typical system can be achieved within 12–60 months. Recent installations of BMS for boiler-optimized controls have delivered savings of 28–54% giving an indication of the energy savings from installing a BEMS by using our energy savings calculator.

The increased drive to energy efficiency coupled with reluctance by organizations to make any capital expenditure has prompted innovative financial models to

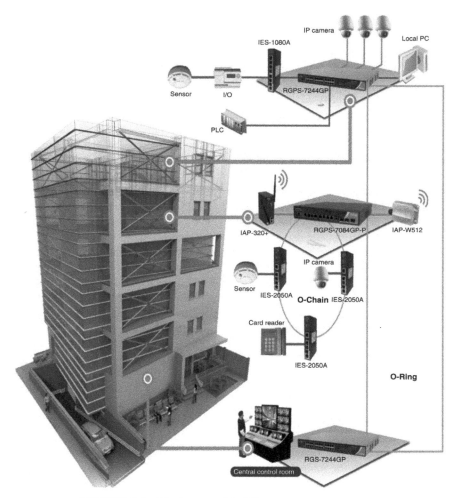

FIGURE 2.5   The integrated building management system.

increase access to new technologies. Guaranteed Savings Contract is an energy performance contract scheme that provides the option of procuring energy controls technology on a "pay-as-you-save" basis. Organizations can then benefit from immediate energy savings – without any capital investment or installation and commissioning costs. As part of the energy performance contract, energy and operational cost savings offset the investment.

The two acronyms tend to be misused to describe the same thing; however, BMS provide a computer-based system that seeks to integrate a comprehensive range of building services. These services can integrate building controls and monitoring, covering such systems as mechanical and electrical equipment, HVAC, lighting, power systems, fire systems, and security systems.

BEMS are generally designed to provide a totally integrated computerized control and monitoring systems of energy-related plant and equipment such as HVAC and lighting, but would not normally provide for the integration of systems such as fire and security as these are not considered to be energy-related systems.

Critical parts of the BEMS hardware and software can deteriorate without regular servicing or maintenance. For example, temperature sensors can deteriorate and fall outside their calibration accuracy or simply become broken or damaged; control dampers and valves can fail to function correctly due to age or wear. The other most common problems occur with temperature dead bands and free cooling options. In addition, control software can become out dated. Ultimately, failure to address problems means that the BEMS no longer serves the intended purpose and is relegated to serving as a glorified time clock to start/stop the energy systems – thus failing to deliver the original intention of energy, CO_2 and cost savings.

Many organizations are wasting energy through inefficient boilers. This is because typical boilers will often fire up and consume energy when it is not necessary to do so – instead of using the residual heat that is already available within the system. This is known as "dry cycling" and can add more than 30% to the cost of an organization's heating bill. Building energy management systems include boiler optimizer controls to ensure that the boiler is demand driven and only operates when required, and only at the temperature required to satisfy demand. This reduces energy consumption and avoids unnecessary wear of the equipment.

2.4 BMS ARCHITECTURES

After introducing the general architecture of a BMS for energy efficiency and briefly describing the main functionalities it should implement, we now survey a number of architectural solutions proposed in the literature, and we analyze and compare them from different viewpoints, such as *architectural model* (e.g., centralized versus distributed), *internal organization* (e.g., single layer versus multilayer), *networking protocols*, ability to support *heterogeneity* in sensing technologies, and so on. Moreover, we compare different solutions with respect to such software quality attributes as *modularity*, *extensibility*, and *interoperability*.

2.4.1 Plain Support for Energy Awareness

The first considered solution is a monitoring system based on *Web-enabled power outlets* [8]. Since the system is only intended to stimulate user awareness to energy consumption, there is no actuation infrastructure. A Web-based user interaction interface is responsible for sending appropriate notification messages to the user. Each appliance is connected to a power outlet, that is, a power meter that measures the energy consumption of the appliance and sends the acquired information to a Gateway, using a standard communication protocol (e.g., Bluetooth or ZigBee). By providing an Application Programming Interface (API), the Gateway seamlessly integrates the smart power outlets into the Web [9]. This allows users to easily access their energy consumption through a Web browser. At the same time, it opens the system to application developers. Such an approach would appear overly simplistic with respect to the ideal BMS; complete focus on energy monitoring does not allow relating consumption to the current environmental state, nor does it allow automatically controlling actuators. A very fine-grained energy monitoring by unintrusive devices would, on the other hand, be advisable for the realization of an ideal BMS, possibly based on a more complex architecture.

In the previous solution, the integration of power outlets with the World Wide Web is mediated through an intermediate gateway. A further evolution consists of a direct integration of power meters, and possibly any other smart device, by exploiting the *Web-of-Things* (WoT) paradigm. The latter is the extension of the well-known *Internet of Things* (IoT) paradigm to the Web [10,11]. Following the WoT approach, any smart object (e.g., power meter, sensor, actuator) hosts a tiny web server. Hence, it can be fully integrated into the Web by reusing and adapting technologies and patterns commonly used for traditional Web content. An application framework for a smart home following the WoT paradigms has been proposed in *Home Web* [11]; this solution is characterized by some degree of modularity because it is based on a Web-service approach. The solutions discussed so far rely on a centralized architecture and are able to support heterogeneous embedded devices, thus providing a basic support for interoperability and extensibility, even if these potential characteristics are not fully exploited.

2.4.2 Integration of Actuators and Environmental Sensors

A centralized architecture consisting of a central server that interacts with heterogeneous sensory and actuator devices is also implemented (Figure 2.6). Specifically, a Wireless Sensor Network (WSN) is used to monitor environmental conditions and to measure energy consumptions, while actuation is performed by X10 [X10 2013] devices connected to the server via power line communication (PLC). Since wireless sensors have a limited transmission range, they may not be able to communicate directly with the server. Hence, to extend the system coverage, sensing devices send their data to a local base station. Base stations are then connected to the server through an Ethernet high-speed LAN.

Rules are defined by the system administrator by means of a high-level language and translated into service requests for the actuators. iPower paves the way for an interoperable, modular, and extensible solution. The iPower solution, despite the adoption of slightly more intrusive sensors and actuators, allows monitoring environmental quantities, besides energy consumption; moreover, a

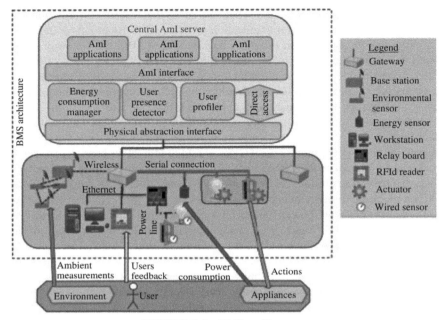

FIGURE 2.6 BMS architecture.

hierarchical organization vouches for medium scalability. However, it is our belief that a greater effort is necessary in terms of scalability, also with respect to the software components devoted to reasoning. The rule-based engine guarantees a coherent source of reasoning, albeit a reactive one, and does not support the prediction. Finally, the actuating infrastructure appears too simple to enact automatic control of actuators, and merely allows tuning their supply power.

A centralized approach, similar to that used in *iPower*, is also considered by *Green-Building* [12]. Unlike *iPower*, *Green Building* uses an unstructured (i.e., single-tier) architecture and combines the energy monitoring and control functionalities into a single infrastructure (i.e., power meters are also actuators). In addition, sensing devices for environmental monitoring can be fully integrated into the same unique wireless infrastructure. A similar solution is also proposed in Ref. [13], where a prototype of a wireless actuation module is presented that can be fully integrated within the monitoring WSN. Using a single (wireless) infrastructure for monitoring and control lessens the burden of technology integration. On the other hand, it reduces the flexibility in deciding the granularity of the monitoring/control process. As for *iPower*, this architectural solution aims at the right direction but does not appear fully adequate yet because of the simple actuating system and the lack of explicit support for intelligent reasoning.

2.5   ENERGY SYSTEMS MONITORING

Energy systems monitoring can be classified according to various criteria, for example, the type of sensors they use, or the spatial granularity used for collecting data. With respect to sensors, it is possible to distinguish between *direct, indirect,* and *hybrid* monitoring systems. Direct monitoring systems use electricity sensors for directly measuring energy consumption, while indirect systems infer energy consumption by measuring other quantities such as temperature and/or noise. Finally, hybrid systems rely on both approaches. Direct monitoring systems can be further classified into *fine-grained, medium-grained,* and *coarse-grained* systems, depending on the level of spatial granularity they use in collecting data about electrical energy consumption. The taxonomy is graphically summarized in Figure 2.7.

2.5.1   Indirect Monitoring

As expected, indirect monitoring systems are so called because they do not use electricity sensors for measuring the energy consumption of appliances. Instead, they indirectly infer information about energy consumption by measuring other physical quantities that are somewhat related to energy consumption.

This approach leverages the fact that appliances typically affect other observable environmental variables, such as temperature, ambient noise, vibrations, or electromagnetic field. Specifically, data provided by sensors are combined with a consumption model of the appliance in order to obtain an estimate of its energy consumption. An indirect monitoring system is proposed in Ref. [14], where a wireless sensor network is used to measure physical quantities such as noise, temperature, and vibrations. Each appliance is identified by a specific pattern of its sensory measurements. For instance, switching on a kettle is associated with the temperature rising, a variation in vibration, and ambient noise. However, the paper does not specify how the system is provided with the association between sensory patterns and specific operating appliance; additionally, the simplicity of this

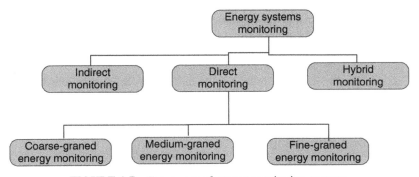

FIGURE 2.7   Taxonomy of energy monitoring systems.

approach limits its applicability to feedback-based systems. Given the use of signature-based models for environmental measurements, this solution could be viable in centralized intelligence architectures, using a distributed sensor infrastructure.

Whenever a model for appliance energy consumption is available, any system capable of automatically detecting appliances could be used for performing indirect energy monitoring. These systems include the approach proposed in Ref. [15], which exploits information coming from the energy distribution network, other than explicit energy consumption. The proposed approach analyzes high-frequency electromagnetic interferences generated by the electronic devices powered through a switch-mode power supply (SMPS) (used in fluorescent lighting and in many electronic devices). Due to the limited applicability to a specific class of actuators, such technology should be just regarded as complementary to the energy monitoring system. For instance, this approach could be suitable for fully centralized architectures where the pervasiveness of sensory devices is minimal.

With reference to the ideal BMS, indirect energy monitoring systems are not suitable because their use would require building models for actuators, which, especially when environmental measurements are involved, would have to be done *in situ*, thus being invasive for users, not well generalizable, and consequently slowing down the deployment of the entire BMS.

2.5.2 Direct Monitoring

Unlike indirect systems, direct monitoring system measures energy consumption through ad hoc electricity sensors, typically referred to as power meters. The granularity used for direct energy monitoring spans from a single point of metering to the monitoring of individual appliances. The rationale for using only a single power meter is keeping intrusiveness at a very low level. These coarse-grained systems are referred to as *NILM* (*non-intrusive load monitoring*) systems, or *NALM* (*non-intrusive application load monitoring*) systems if the focus is on individual appliances.

On the opposite end, fine-grained systems allow monitoring individual appliances with a high precision but require the deployment of a large number of power meters. Obviously, the granularity of monitoring affects the approach to the artificial reasoning carried on the collected sensory data and, indirectly, also the possible energy-saving policies that can be used.

Such detailed monitoring, not available in the approaches mentioned so far, is useful to avoid using consumption models for those appliances, thus eliminating the initial training phase with its costs in terms of user discomfort.

2.5.3 Hybrid Monitoring

Finally, a hybrid approach to monitoring, including both direct and indirect parts, involves using both specific sensors for energy measurement (typically in a single power meter at the root of the distribution tree) and indirect sensors for recognizing the operating status of appliances. A monitoring system based on WSNs with magnetic, light, and noise sensors and including a power meter is used for monitoring the overall energy consumption. An automated calibration method integrates two types of models for learning the combination of appliances that best fits the collected sensory data and the global consumption. Specifically, a model of the influence of magnetic field, depending on two *a priori* unknown calibration parameters, is used for more complex appliances with many operating modes. On the contrary, appliances with fewer operating modes only require models associating the relative consumption to each specific mode, which is estimated via the noise and light sensors. The main disadvantage of this work is that the calibration is to be performed *in situ* and cannot be carried out before the deployment because many unpredictable external factors may influence the measured environmental variables. It is worth pointing out that hybrid systems are typically characterized by a coarse-grained direct monitoring of energy, with a single sensor at the root of the energy distribution tree. This is usually coupled with a fine-grained indirect monitoring.

2.5.4 Comparison of Different Energy Monitoring Systems

An ideal BMS that is able to provide an accurate description for actuator consumption without demanding excessively intrusive deployment naturally calls for fine-grained direct monitoring. However, when deployment costs are prohibitive, it is possible to reduce the number of used devices and to rely on a disaggregation technique, starting from the branches of the energy distribution tree.

Figure 2.7 reports a comparison of different energy monitoring systems together with some of the previously discussed architectural solutions according to two qualitative dimensions, namely, the overall intrusiveness experienced by users and the details on attainable monitoring. Values along the first dimension were attributed to assess both the intrusiveness of deployed devices and the discomfort perceived by the users during the training phase, while the second dimension is tightly related to the position of the assessed solutions within the taxonomy depicted in Figure 2.7. Note that, as regard the sensory infrastructure, costs get higher as the systems get closer to the ideal one. When it is important to keep installation costs below a given threshold, it will be necessary to trade part of the functionalities of the final BMS for cost.

2.5.5 Devices for Energy Sensing

Besides the different approaches to energy monitoring, it is also necessary to consider the technology to be used for creating the sensory infrastructure. In this section, we will mainly focus on the available technologies for energy sensing. However, we will not consider sensors for environmental monitoring, as they are beyond the scope of this survey, although they are exploited into indirect and hybrid monitoring systems. A wide selection of sensor technologies for energy sensing is currently available on the shelf. The choice of a given technology directly affects the complexity of the architecture supporting the monitoring system and providing the integration with the rest of the BMS. Energy consumption models of individual appliances represent an alternative tool for energy monitoring, as they allow estimating the overall energy consumption of buildings simply relying on the knowledge about the status of each appliance (i.e., without requiring any specific sensing infrastructure). For some devices, the corresponding energy consumption in different operating modes can be retrieved from their technical specifications, such as the Code of Conduct (CoC) edited by the European Commission, which, however, do not cover the entire set of available appliances.

2.5.6 Integrated Control of Active and Passive Heating, Cooling, Lighting, Shading, and Ventilation Systems

Buildings account for nearly 40% of global energy consumption [7]. Of this, about 40 and 15% are consumed by HVAC and lighting systems, respectively. In view of the increasing energy cost, government mandates for energy efficiency [8], and the rising human comfort requirements, controlling shading blinds and natural ventilation to make effective use of natural resources can reduce energy consumption and is therefore of great interest [9,10]. In addition, improving the HVAC control can also result in significant cost savings [11]. HVACs, lights, shading blinds, and natural ventilation interact with each other in energy consumption via thermal phenomena and in satisfying human comfort requirements for temperature, humidity, fresh air quantified by CO concentration, and illuminance in each room. As shown in Figure 2.8, the indoor temperature is affected by all the above-mentioned devices; both indoor humidity and CO concentration are affected by HVAC and natural ventilation, and illuminance by lights and shading blinds. In summer, for example, if blinds are open for using the daylight, energy consumption of lights is reduced. However, energy consumed by HVAC will increase due to the increased solar heat brought by inlet sunlight [9]. Therefore, the control of

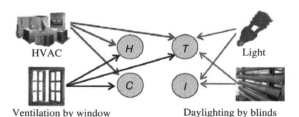

FIGURE 2.8 Couplings of different devices on human comfort. T: temperature, H: humidity; I: illuminance; C: CO concentration.

blinds must consider not only the energy consumption of lights but also that of HVAC. Integrated control of these devices is important to manage such inter-actions. In addition, individual rooms share an HVAC system and are coupled in competing for its limited capacity. Integrated control of these devices is therefore also important for preventing the cooling demand from exceeding HVAC capacity and essential for human comfort [11]. In most of the buildings, active and passive sources of heating, cooling, lighting, shading, and ventilation, however, are not coordinated. Analytical studies on their optimal integrated control have not been found in the literature. Possible reasons might be that (i) it is difficult to establish models that have a good balance between accuracy and simplicity for optimization; (ii) models are difficult to calibrate [16]; and (iii) the interactions between devices and the coupling among rooms make it time-consuming to search for the optimal or effective control strategy.

2.5.7   Electricity Network Architectures

Traditional electrical power system architectures reflect historical strategic policy drivers for building large-scale, centralized, thermal (hydrocarbon and nuclear)-based power stations providing bulk energy supplies for loading centers through integrated electricity transmission (high voltage: 400, 275, and 132 kV) and distribution (medium, low voltage: 33, 11, 3.3 and 440 V) three-phase systems. In the mature economies, these designs have been predominant, but as a result of industry restructuring and international policy drivers for low-carbon renewable energy production, they have been underinvested and are now in question as to their future sustainability with regard to anticipated future energy scenarios that may compromise their ability to support innovation. The hierarchical control structures for these traditional designs differ across the transmission and distribu-tion levels with greater automation (and complexity) obvious at the high-voltage levels, with centralized control-room-based operational management and reason-ably pervasive communications capabilities for automatic control and system protection. At the distribution level, conventional network design has led to less sophisticated system control and management structures with lower levels of automation in place. Figure 2.9 indicates the high-level changes emerging in electrical power system architectures in response to managing aging assets and increasing levels of distributed generation connections. Given the significant growth and penetration of renewable sources and other forms of distributed generation, there are now increasing pressures on distribution networks to cope with new system stability (voltage, transient, and dynamic), power quality, and network operational challenges brought about by embedding generation sources that would have been, more typically, larger scale, thermal, and connected to the grid at the transmission levels. Consequently, we are approaching a problem inversion situation where, similar to conventional transmission networks, more active network strategies and technologies will be required at the distribution level. Figure 2.9 presents a number of issues in network management and the

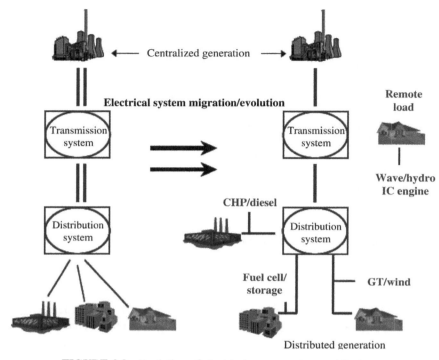

FIGURE 2.9 Evolution of electrical power system architectures.

resulting changes to conventional methods of system control. The term "active" is significant because the medium-voltage distribution network (unlike the high-voltage transmission net- work) has traditionally been a passive means to pass power from bulk supply points to customers. The quality of supply has been ensured by planning a degree of redundancy and by some centralized ability to switch connection points. Single-circuit radial distribution lines are vulnerable to faults and the first priority of power system protection schemes is to isolate faulted sections and plant. Restoration of customers that are off supply can be a relatively lengthy process because automated restoration relies on methods run by controllers that are written for only a small number of scenarios. If the scenarios do not apply, then restoration is through manual control. Voltage profiles in the network are assessed at the planning stage and transformer tap-changers (perhaps with line-drop compensation) used to accommodate load variations.

The inclusion of distributed generation sources calls for a greater degree of control, including control of distributed generation reactive power. It is therefore not straightforward to integrate new distributed generation and effect its connection to the network. Without active network management, the full network capacity

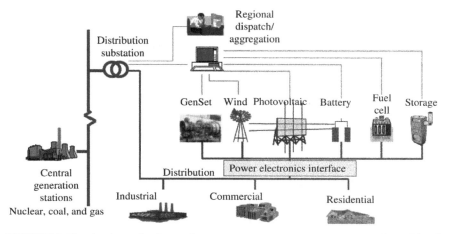

FIGURE 2.10   A schematic of a semiautonomous power system or power cell capable of managed islanding from the main grid.

potential cannot be realized. Active network management is about the integration of distributed generation into network control with greater coordination of power system operation, rather than its straightforward connection. Active network management can also make use of other distributed resource, such as storage, to relieve constraints that arise in networks where energy use and demand patterns have changed. Technical analyses have demonstrated that by employing active network management methods, distribution networks can accommodate about three times more distributed generation connections than equivalent networks without active management (Figure 2.10).

2.6   ENERGY SAVINGS FROM BUILDING ENERGY MANAGEMENT SYSTEMS

Building energy management systems have the ability to save energy and improve productivity by creating a comfortable working environment. Our world is currently facing two particularly important trends: rising fossil fuel prices and concerns about climate change. Both create strong incentives for energy conservation.

The World Business Council for Sustainable Development identified buildings as one of the five main energy users, where "mega-trends" are needed to transform energy efficiency. Buildings account for 40% of primary energy in most countries and consumption is rising. The International Energy Agency (IEA) estimates that for buildings, current trends in energy demand will stimulate approximately half the energy supply investments through 2030.

Building energy management systems have the ability to save energy and improve productivity by creating a comfortable working environment. BEMS optimization create improved energy management; however, regular building audits and fine-tuning are necessary to ensure the energy management is maintained. The technical strategies for achieving energy savings are summarized while optimizing occupant comfort. BEMS optimization is dependent on the physical plant, operator, level of controls, and zoning, as well as the type of environment to which the system is being applied. This information is targeted to internal energy savings implementation professionals, looking for a resource to guide them in changing parameters, tuning building management systems, and recommissioning existing systems.

2.6.1 Energy Savings Opportunities

The easiest way to create savings is to reappraise and/or relax set points. Caution must be applied as the changes need to be made in accordance with the overall building scheme, as the settings may be a crucial part of an overall control strategy. A shift can be applied in accordance with external conditions; for example, with an air-conditioned building, the summer set point for cooling can be increased relative to an increase in outside temperature (within a predefined band).

A regular review of set points and modification is an essential part of the ongoing energy cycle and must be continually reviewed, looking for opportunities for further savings. When an opportunity for set point savings has been identified, minor set point changes over a period of time ensure a smooth transition; for example, a stepped changes of 0.5 °C or 1 °F at a time for room temperature.

Shifting/relaxing set points in line with a combination of external conditions and time/calendar rationalization can typically equate to 5–20% savings. A set point reduction can equate to 10% savings per degree on your heating bill, with potentially higher savings on cooling/chiller bills.

Occupancy-Time Schedule Ensuring your building operates according to occupancy levels is a key energy saving action and requires continuous reviewing to ensure the settings are representative. For example, occupancy patterns of schools and universities continually change due to activities such as after-school school clubs, evening classes, and so on. It would be easy to apply a "carte blanche" approach, setting a broad range time pattern, but this would equate to unnecessary periods of heating and cooling. Regular review of occupancy levels would highlight the possibility to change set points for multiple periods of occupancy on different days. In addition to a permanent change, the ability to extend a time operation on a one-shot basis or on a 0–30 min timer ensures that a one-off change in occupancy, such as would occur with an unexpected late meeting, is changed for that one period and then revert to the normal occupancy pattern.

Zoning A cost-effective way to save additional energy is to apply further zoning to areas where there are different occupancy patterns. These zoned areas are only heated or cooled when required. Each zone can have occupancy times, compensation, and optimization applied to maximize the savings potential.

Calendar Schedules   BEMS offer advanced time scheduling capabilities, and within this is the ability to apply schedule patterns for different calendar dates. This enables variable time scheduling to match varying work patterns to be programmed well in advance. This option can be applied to areas where occupancy levels are constantly changing week to week, such as exhibition halls or meeting rooms. Operator time is thus reduced because configurations are made once as opposed to making changes on a weekly basis.

Holiday/Vacation Periods   To ensure energy savings during public holidays when businesses are closed, holiday schedules are used in conjunction with time schedules. For example, In the United Kingdom, typically there are eight public holidays. To determine the energy savings for a commercial property, multiply the facility availability of 52 weeks by 5 working days = 260; therefore, eight public holidays equates to over 3% possible energy savings.

With an integrated systems approach, a single change to a core time schedule or holiday schedule can propagate to all integrated systems, including lighting, security, and access control. This ensures HVAC systems work in empathy with the actual required occupancy, therefore maximizing energy savings throughout the building by reducing operating costs.

Optimizers   Prior to the introduction of optimizers in the mid-1970s, many buildings were controlled entirely by a mechanical time clock. These were often set to switch on the building at a specific time and often assumed the worst weather conditions, such as heavy snowfall, thus running the building's central heating system from the early hours of the morning till the late evening, without change.

Synonymous with energy savings is the "Optimizer." Prior to the introduction of the BEMS, an optimizer was a stand-alone controller with an outside temperature sensor located on a north wall and internal space temperature sensor(s). A temperature rise rate was calculated in accordance with how cold it was outside and this became a time factor that was tuned based on the heat loss of the building and the difference between the internal temperature and the desired occupancy temperature.

Based on this, the plant was switched on at a time prior to the required occupancy time, which equated to putting in the "optimal" amount of energy. The start time depended on the external temperature, the indoor temperature, and how much energy was required to meet the desired occupancy space temperature at start time.

The optimum "off" function that worked the opposite way was the next innovation to follow. It predicted the "off time" based on the external temperature, the room temperature, and the earliest possible time the building could have its

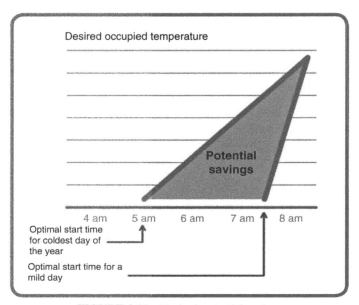

FIGURE 2.11   Desired potential saving.

heating plant switched off, while still retaining comfort conditions at the end of the occupancy period.

A low-temperature protection setting is applied to protect the internal fabric of the building that can be damaged through condensation should the temperature/humidity condition reach dew point. Optimizers provided typical energy savings of 5–25% (potentially higher with cooling/chiller plant) compared with standard controllers where a limit of 2 h is applied to the start-up time (Figure 2.11).

The BEMS provides extensive reports on the optimizers' operations and they must be regularly reviewed to ensure the maximum savings are achieved. This can be done after different external temperature conditions and on different days of the week. The BEMS optimizer has additional "boost" functions that may be applied if the internal temperature did not reach occupancy levels in the previous 24 h, such as would be the case on a Monday morning. This is enabled automatically to ensure comfort levels are achieved.

Frost Protection   It is fundamental that when a building is switched off either in normal operation or in holiday/vacation mode, a frost protection strategy is in place. Frost protection strategies will allow pumps and the heating system to remain off when the building is not occupied to save energy. The pumps and heating system will energize when the temperature outside, in the main pipework or in the space, fall outside of acceptable ranges.

Overrides   In instances where systems are occasionally manually overridden, a regular review or identification is essential to ensure energy is not used unnecessarily.

Compensation   With a water-based system, such as radiators, compensation is normally applied whereby the temperature in the circuit varies in accordance with the external temperature. The more the colder outside, the higher the water temperature in the circuit. There is a minimum and a maximum setting applied. This must be reviewed regularly or after any overhaul to ensure the compensation parameters are still representative and prevent overheating, which typically saves 5–10% on energy use.

Standard compensation can be enhanced by the addition of room influence, solar influence, and wind influence, whereby a number of sensors are fed back into the control loop and influence the set point. This, in turn, provides improved comfort conditions and prevents overheating. It is important to ensure that the maximum ΔT (temperature difference) for your system is achieved/maintained for any boost period to ensure the quickest consistent run-up and boiler efficiency.

Outside High Limit   A water-based heating system, even with compensation applied, can be switched off if the outdoor temperature exceeds a preset value where the difference between internal and external temperatures is minimal or even negative. Heating is not normally required in a building when the outdoor temperature exceeds 16 °C or 61 °F, depending on the building type. It is important that hysteresis is applied to prevent plant turning on and off rapidly with a minor temperature change outside. Hysteresis is a method of control that will keep the plant turned off until the temperature rises a few degrees above the set point – similar to a household thermostat. Each building is different and the set point should be calculated accordingly.

A low limit can be applied with cooling to ensure free cooling is used when the external temperature is below a preset value by closing a cooling valve, zone, or disabling the primary chilled water plant (see "Enthalpy Control" section). For example, a chilled water plant is disabled when the external temperature falls beneath 12–14 °C or 54–57 °F. Providing cooling is not required for process or there are no significant heat sources within the building.

Disable Humidification   If the humidity (outdoor moisture content) is above the required level and satisfactory humidity levels are achieved in the return duct, then humidification systems can often be disabled. This application must be reviewed on an individual air-handling unit basis to ensure the control scheme allows this. Some air handlers rely on 100% humidified air to reheat the supply to the desired level. Location of people and equipment is a consideration.

Control Stability   A lack of stable control increases energy usage by typically 3–5%, and decreases the life of valves and actuators. Primary heating, chilled water, and central air-handling units must provide a stable supply temperature to their served areas, such as distributed air-handling units, VAV boxes, or fan coil units. Unstable primary plant and/or the local plant control having incorrect PID settings cause hunting. Hunting occurs when a system first overcorrects itself in one direction and then overcorrects itself in the opposite direction and does not settle into a stable position. Figure 2.11 shows a graph of unstable control where the supply temperature increases and then decreases continually. This can cause overheating followed by overcooling, which may only equate to a slight +/− variation around the temperature set point, but causes mechanical wear and tear, as well as inefficient energy usage.

By physically watching the control items for movement, the BEMS' trend analysis capability monitors valve positions and assists in the fine-tuning of the control loop to maximize savings. Unstable control can occur due to changing plant performances and efficiencies. For example, a blocked filter reduces airflow. Regular reviewing of control loop performance is important to highlight failing loops or those that are hunting.

Air-Handling Systems: Damper Economy Override   Most air-handling unit systems consist of a supply and extract with a recirculation duct with dampers on each side to recirculate the already heated or air-conditioned return air or to utilize fresh air as a free cooling source. Fresh air brought into the building is usually set to a fixed percentage (typically 10%). By using an air quality sensor in the return duct, the percentage of fresh air can be reduced when air quality is good, which is normally at the beginning of a working day, equating to energy savings and increased occupant productivity. Variable air volume systems need to maintain air by volume that can be used in conjunction with air quality.

Enthalpy Control   Enthalpy is the total heat content of the air. This can be applied to air-handling unit systems with heating and cooling and humidity control. The principle is that even though the outside air may be warmer than the return air, there can be less total heat in kJ/kg of energy. A software algorithm is used to set this switch and dampers are positioned to utilize the "warmer" outside air that has a lower total heat content (Figure 2.12).

Demand Programming   This program will constantly look at the heating and cooling control valve positions to determine if there is a load on its associated system. If any (or a low percentage) of the valves are open more than 5%, then the systems operate normally to satisfy the demands. If all the valves (or a high percentage) are less than 5% open, then the secondary pumps are disabled. After a time delay, the primary pumps and main heating or cooling systems are disabled, provided there are no other load demands from any other systems. This improves the efficiency of the primary system, as it only operates during a predefined time schedule if there is a genuine demand and not just because the time schedule is on.

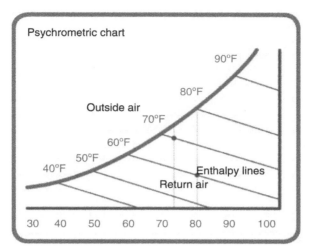

FIGURE 2.12   Psychrometric chart.

Night Purge/Summer Precooling   If the cooling load at the start of building occupancy is required and if the nighttime external air is cooler than the required occupancy temperature, then night purge can be applied. This sequence enables central heating and chilled water plants to be disabled and air-handling unit systems to run in full fresh air mode for a period of time, typically 30 min, in the early morning hours, before the sun has risen. This fills the building with fresh, cool air and reduces the initial load on the primary system at occupancy start. Flushing the building with fresh air also clears out residual carbon dioxide/vitiated air and provides building occupants with cleaner air.

Electricity Savings: Load Cycling   Load cycling refers to switching off an electrical load for a period of time on a regular basis. Load cycling can be applied to background systems, such as a fan or pump so that it will not result in consequential inconvenience. You should override load cycling if conditions exceed a preset value, such as a low space temperature. For example, if the system is switched off for 5 min within a 20 min period, then the savings per hour equals 20 min or 25%. When applied, load cycling typically results in 5–25% savings on the electricity bill, depending on the size of the plant.

   Disadvantages of load cycling are that regularly starting and stopping plant may cause an increase in electrical load during start-up and could decrease the overall life of the plant. In these cases, the use of variable speed drives should be considered (Figure 2.13).

Variable Speed Drives   The use of variable speed drives in various aspects of a building is now prevalent. Many are used mainly as soft start-up and then operate at a fixed speed. The information held within the BEMS can relate to

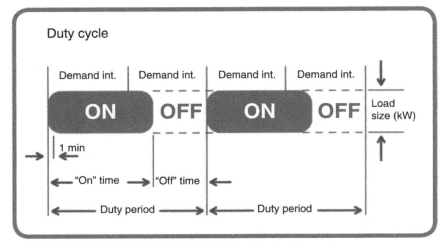

FIGURE 2.13   Load cycling for electricity saving.

environmental conditions and occupancy levels from access control, with these data algorithms relating to demand. For example, varying the air volume through the working day, based on occupancy levels from the access control or air quality sensors, ensures that the minimum amount of energy is used on any partially occupied area of the building. Reducing a 50 Hz motor by 20% to 40 Hz equates to 50% energy reduction.

Maximum Demand   Maximum demand sets a limit for the maximum consumption allowed (normally over a 30 min period) and is a cost-reduction measure by preventing this limit from being exceeded. If anyone exceeds the limit, then a "penalty" is applied to the electricity bill that could equate to paying a higher tariff per kW/h consumed. Therefore, the aim is to ensure that the maximum demand limit is not exceeded. Cost reduction associated with maximum demand implementation can be substantial if demands were regularly exceeded and penalties applied.

A controller is synchronized with the maximum demand meter and forecasts whether the limit will be exceeded by monitoring the rate of electricity consumption versus the amount of remaining energy and time. The algorithm associated with maximum demand is complicated, but the net result is that site-wide electrical loads are shed if the algorithm predicts the limit will be exceeded. Electrical loads are reinstated after the danger period has passed. Electrical loads are shed in rotation per priority level and a matrix enables the choice of load criticality. The rate at which they are shed and restored is continually reviewed by the calculations.

The required demand target can be determined using the BEMS if further reductions in electricity consumption are required. Determining which electrical

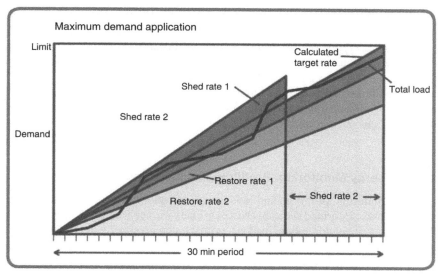

FIGURE 2.14   Maximum demand reduction.

loads can be shed can be complicated. The lowest level may be electrical water heaters, the highest level may be one of a number of chillers, whereby it may be out of sequence for a period of time as it goes through a shutdown sequence before it is reintroduced to the control scheme. Indirect reduction of maximum demand could be applied by overriding the amount that a chilled water control valve can open to. This would indirectly reduce the load to the chilled water plant and, therefore, reduce electricity consumption; however, the time the valve takes to do so may not be practical, but may be possible on parallel routines. The maximum demand reduction is shown in Figure 2.14.

2.6.2   The Intelligent Building Approach

Intelligent integrated building solutions are becoming standard. Building integration can include access control, intruder detection, security, chillers, lighting, digital video, power measurement, variable speed drives, and so on. The integrated approach provides access to all building systems through one coherent and customizable user interface. Additionally, building integration reduces training costs and standardizes alarms and logged data.

Integrated building systems also lower capital expenditures because data networks are shared, there are fewer computers and servers, and devices have numerous uses. For example, a passive infrared detector, normally only used by the intruder systems, can also trigger CCTV recording, relax set points for HVAC control, and turn off lighting when no occupancy is detected. Another example, when access control is used to gain entry to a building, this signal is used by the lighting control and HVAC systems to change from economy levels to occupied mode.

Ongoing operating expenses are also reduced because there are fewer computers and networks to maintain and fewer user interfaces, ensuring those who operate them are more efficient and productive. Integrated control strategies offer extended energy savings by allowing the building systems to work in empathy with each other. Using information from all the systems, strategies can be deployed to reduce the use of energy-consuming devices and create a comfortable and productive workspace.

2.6.3 Energy Monitoring, Profiling, and Modeling

Energy monitoring, profiling, and modeling applications provide the information needed to make informed decisions based on energy usage patterns. Understanding and reducing the building base load is a primary step in reducing utility costs. The required data for energy monitoring application is shown in Figure 2.15. The system data can be gathered in intervals (15, 30, 60 min) by the electricity utilities data provider (mandatory where consumption exceeds 100 kW/h in some markets). Gas and water meters are often connected to spare inputs. With the customer's approval, utility grade data are accessed along with the hardwired or soft-calculated BEMS meters and are further processed to enable presentation and data analysis through a secure Internet site. This information can be graphed throughout the day allowing you to see energy use rise when building systems start and energy use decrease when occupancy and building use decreases.

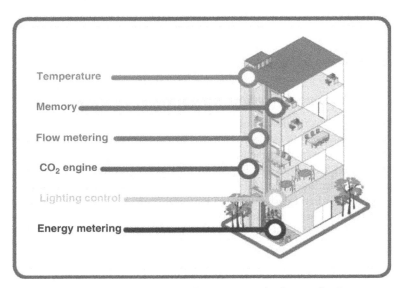

FIGURE 2.15 Data required in energy monitoring applications.

This information is used to validate energy consumption; for example, you can ensure energy consumption matches the actual occupancy of a building, taking into account any preheat/cool cycles. The load profile is the focus for energy optimization to (a) understand and optimize the building base load consumption, (b) reduce peaks, and (c) reduce daytime use. The ability to compare and benchmark information by overlaying equivalent days such as a Monday's profile or a specific week's profile provides an accurate picture and highlights anomalies for investigation.

Modeling enables "what-if" scenarios to run on existing data factors. For example, "What if I reduce my energy by 10% between 09:00 and 11:30, or by 16 kW between 17.30 and 19.59?" with visual feedback in terms of energy reduction, CO_2, carbon, and so on. The utility modeling cost-reduction techniques can deliver savings for various industries. The modeling tool is easy to use and is provided as a web service on a day +1 basis.

Utility performance visibility complements the real-time alarm and controls facilities of the BEMS software. Importantly, it extends the benefits of a single utility meter, as meters can be soft calculated for smaller areas of the building, giving additional perspectives of the site's performance, such as a consumption profile for a given department or tenant, as well as trends and savings achieved through investments and so on.

Energy Aggregation   Energy aggregation is used when there is more than one site involved. The use of technology can collect, aggregate, and analyze total energy usage and more importantly the overall consumption profile, spanning all buildings (Figure 2.16). The data can be used to negotiate improved tariffs; based on the aggregated profile, significant savings can be negotiated.

2.7   SMART HOMES

A *smart home* may be defined as a well-designed structure with sufficient access to assets, communication, controls, data, and information technologies for enhancing the occupants' quality of life through comfort, convenience, reduced costs, and increased connectivity [12]. The idea has been widely acknowledged for decades, but few people have ever seen a smart home, and fewer still have occupied one. A commonly cited reason for this slow growth has been the exorbitant cost associated with upgrading existing building stock to include "smart" technologies such as network-connected appliances. However, consumers have historically been willing to incur significant costs for new communication technologies, such as cellular telephones, broadband Internet connections, and television services.

A home is already a well-designed connector for power transfer between the electricity grid and energy-consuming appliances. A *smart home* also functions as a switchboard for data flow among appliances and participants such as the end-user, the electric utility, and a third-party aggregator [17,18]. This evolved

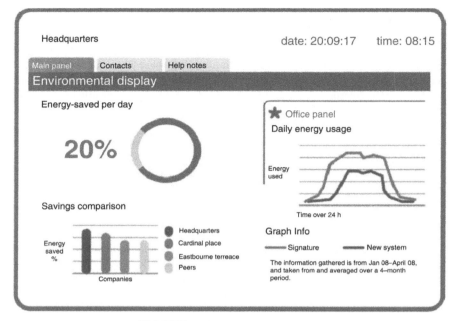

FIGURE 2.16   Total energy usage.

capability benefits stakeholders on both sides of the interface – utility customers, utilities, and third-party energy management firms – because there are strong incentives for all sides to help the others function smoothly. For instance, a homeowner may not inherently care about the peak demand issues faced by the utility, but electricity prices and supply reliability are tied to operational practices of the service provider. On the other hand, a utility may be primarily concerned with meeting the requirements of public utility commissions, but unhappy rate payers may result in business and regulatory risks. Looking outward, a smart residential building has two-way communication with the utility grid, enabled by a smart meter, so that it can interact dynamically with the grid system, receiving signals from the service provider and responding to information on usage and diagnostics. This bidirectional information exchange is enabled by the rapid adoption of advanced metering infrastructure (AMI).

Looking inward, a smart home employs automated home energy management (AHEM), an elegant network that self-manages end-use systems based on information flowing from the occupants and the smart meter. The value of AHEM is in reconciliation of the energy use of connected systems in a house with the occupant's objectives of comfort and cost as well as the information received from the service provider. Sensors and controls work together via a wireless home area network (HAN) to gather relevant data [11], process the information using effective algorithms, and implement control strategies that simultaneously co-optimize several objectives: comfort and convenience at minimal cost to the occupant, efficiency in energy consumption, and timely response to the request of the service provider. An example of a smart home is constructed in a laboratory setting at NREL [16].

2.7.1 Economic Feasibility and Likelihood of Widespread Adoption

Several market and technology trends are expected to accelerate the development of cost-effective AHEM systems that enable smart homes. These include the following:

- Implementation of smart grids and continued growth in home offices will expand market penetration of secure HANs.
- Growth in web-based cloud computing applications will enable low-cost home energy data storage, data display, and data analysis for AHEM trend analysis [19].
- Advancements in smartphone technology such as batteries, user interfaces, and material [12] are expected to aid the development and adoption of AHEM systems.
- Manufacturers of residential equipment and appliances continue to embed additional sensors and control capabilities in new, smart home appliances that are Internet-ready, can respond to requests from service providers, and offer advanced cycle controls such as multimode or variable speed controls and fault diagnostic sensors for space-conditioning equipment and "eco" modes for dishwashers, clothes washers, and other major appliances [13].
- Integration of energy services into other networked product offerings, such as security systems and television and telephony service.

A key strategy to engaging all stakeholders may lie in changes to the end-user electricity pricing structures – from fixed tariffs to dynamic prices that may change several times over a day – that reflect the use of the assets on the grid at any given time. If these structures are implemented to provide a tangible financial incentive for customers to respond to the requests of the service providers for demand reduction, the customers can receive measurable monetary value for their participation, in addition to the increased reliability of their service. Financial incentives are but one motivating factor for the adoption of smart homes.

2.7.2 Smart Home Energy Management

Large-scale demonstration efforts have thus far approached smart home research with a strong utility focus and less homeowner focus. Currently, the incentive for homeowner participation is limited to relatively small financial gain via utility pricing structures; otherwise, the motivation is primarily altruistic (i.e., environmental benefits). Most utilities offer incentives for energy upgrades and many have leveraged load-shedding technologies that cycle air conditioners during peak load events. Increasingly, utilities are funding more elegant efforts for on-request load reduction in the residential sector to demonstrate a load reduction system that can alter air conditioner and water heater set points and pool pump operation at the end-user facility during peak load times to enable substantial peak savings with limited impact on their customers [18]. Some utilities provide near-real-time data to homeowners, along with several pricing structures and load reduction requests [20]. Many companies have recently incorporated web-based user interfaces, so a homeowner can adjust thermostat settings or turn off lights from a smartphone, or a web browser [13].

Advanced grid measurements using AMI infrastructure are being rolled out in some utilities [21]. These projects have a multipronged focus on better integration of renewables, enhancement of efficiency, and optimization of consumer demands with utility needs on a community scale. Emerging nonintrusive load measurement systems can provide enabling data, but these modern measurement techniques are not yet robust, accurate, easy to install, or cost-effective for integration at the meter [22]. The available legacy methods for load disaggregation use algorithms supplemented with estimation; so the results may have less relevance to a given household than across an aggregated population [22].

2.7.3 Assets and Controls

In smart homes, many loads can be considered as assets that can participate in the efficient use of electric energy: thermal loads, electric vehicles, and smart appliances. By intelligently controlling their behavior in either a reactive or a coordinated manner, these assets can provide leverage for energy and cost savings [23]. Thermal loads, such as air-conditioning, electric space heating, and water heating, can be controlled by "intelligent" thermostats. Contrary to traditional thermostats operating according to the hysteresis principle, an advanced thermostat such as the Nest has a learning capability that can automatically learn from user behavior patterns [24]. Then, the thermostat adapts the room temperature efficiently, for example, by autoscheduling heating according to arrival and departure times and by detecting when the users are away [24,25]. These strategies can help reduce energy consumption, especially when traditional or programmable thermostats are not configured properly, or cannot detect that users are away. Detailed control of household loads would allow the inherent thermal inertia of smart housing stock to be used for energy storage. The controller could "learn" the thermal response of the home, including factors such as weather forecasts, weather observations, and load levels from monitored devices. The resulting model would better predict future loads, which could be used locally or aggregated for the utility to plan short-term control options. For example, a smart home controller could precool a house in the morning, before the system peak load, reducing air-conditioning loads when signaled from the utility.

Plug-in electric vehicles, including hybrids, are expected to represent 1.7–3.5% of all US light duty vehicles by 2025 [26]. These correspond to a significant domestic load interfaced with power electronics that can also help make homes smarter. Using the vehicle-to-home technology, they can temporarily power the household, for example, during demand peaks when power may become more expensive and the battery can provide a part of the total demand, or during outages by powering the entire household until the battery reaches its lower state-of-charge threshold [27,28]. Adapting the charging schedule according to grid supply conditions offers additional possibilities. The utility of such distributed storage may be improved when used together with distributed generation sources, such as photovoltaic panels [27].

A growing number of domestic loads use DC power internally, including electronics, solid-state lighting, and variable-speed motors. Most small, distributed, renewable energy sources generate DC power, which must be converted to AC for grid connection. Some recent work has considered household-sized distributed storage systems for local backup power and ancillary service provision [29,30]. The convergence of these sources and loads provides an interesting opportunity for significant advances in the granular control of loads and high penetration of small-rated DC-powered assets.

A smart home could also integrate a low-voltage DC bus. Renewable resources, battery storage, and potentially vehicle charging could all interconnect on a DC bus. The DC bus would be integrated at a single point, and many inverters and converters would be reduced to DC–DC converters. When high volumes drive down costs, this simplification could reduce the cost and improve the efficiency of renewable systems, solid-state lighting, and electronic loads. However, this paradigm shift presents challenges in electrical protection, rewiring, and standardization. At present, standards for DC distribution and usage are being developed, including 24 and 380 V distribution systems [31]. Use of DC power distribution remains a retrofit challenge for existing US housing stock, but researchers are studying combined AC/DC distribution using existing building wiring [32].

Appliances also hold potential for smarter energy use. Dishwashers, washing machines, and clothes dryers can be scheduled in advance, and do not need to be directly controlled by the user. The starting time can be postponed by several hours, with no impact on the user as long as the cycle is over when the user requested it initially. A similar strategy can be used to control freezer and refrigerator cycle so as to reduce peak demand by coordinating their operation [30]. Finally, many other loads can provide resources for smart energy use and increase the comfort of the user, including automatic blinds that adjust based on daylight intensity, adaptive lighting, and autonomous vacuum cleaning robots. These devices exploit the possibilities offered by the extensive use of sensors, sometimes forming wireless sensor networks, and actuators controlled by smart, adaptive, and possibly learning algorithms.

Almost all loads are, or could be, equipped with intelligent controllers, ranging from simple on/off control of state lighting to sophisticated controllers for

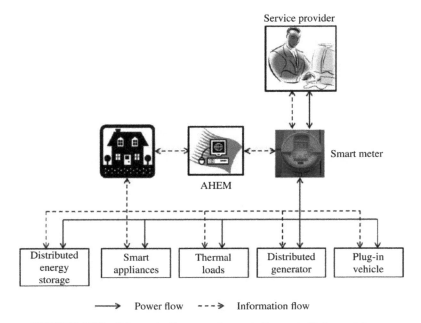

FIGURE 2.17 Schematic diagram of a centrally controlled smart home.

photovoltaic systems, vehicle chargers, and large loads such as air-conditioning. With appropriate standardization and high volumes, practical, low-cost communication systems could connect most loads to a central household controller. The controller could provide detailed monitoring and control for occupants. With proper AMI interfacing, the home could further aggregate the resources for system users, requested by the service provider. A block diagram of the centrally controlled smart home and its constituent assets is presented in Figure 2.17.

If properly designed, controllers could also monitor loads and identify system issues, such as unexpected increases in power draw, current harmonics, or vibration. Significant value – economic and personal – could be derived from identifying issues in advance of catastrophic failure. For many utilities, a smart meter constitutes a smart grid. For others, these smart meters can be put to greater use and provide more substantial value to the utility, the grid, and the end-users via coordination. Analogously, smart homes may span the spectrum from the simple addition of discrete features – such as smart appliances or remotely controllable lighting and thermostats – to an automatically controlled, highly coordinated self-learning system with grid interaction. In the latter case, the control system serves as the brain of the smart home by automating domestic chores and providing sufficient feedback and communication. This symbiotic relationship improves the user's quality of life and allows active participation in bulk power system operations.

There are two schools of thought about the overall purpose of the smart home control system. The first school of thought posits that an ideal smart home control system should be entirely automated, predicting a user's every whim and reacting accordingly so as to maintain user-centered optimal comfort, convenience, and if applicable, savings [14]. One of the tenets of this prevailing theory envisions minimal user input. The control system may incorporate a machine learning algorithm to predict a user's desires as they occur. The second – and competing – school of thought envisions smart homes with well-informed and engaged users that value energy sustainability and are thus active participants in the everyday electricity management of the home [20]. In this case, the consumer is enabled with timely feedback on costs, energy, and emissions to influence the appropriate control strategy.

Machine learning, rule-based, multiagent, and decision-making systems [14] constitute the state of the art in control strategy paradigms for the smart home. Although several smart home control systems are commercially available, they are currently cost-prohibitive to the average consumer; these are expected to become affordable as enabling technologies mature.

2.8   ENERGY SAVING IN SMART HOME

Smart home products not only make your life safer, more convenient, and more fun, they can also help you to save energy and money. As a member of the Flex Your Power energy efficiency campaign, we can show you how to be a friend to the environment and your wallet through energy conservation. The average home spends almost $2000 on energy costs every year. Lower your energy bills and improve comfort by making your home more energy efficient. The average household could cut a third of its current energy bill by switching to energy-efficient appliances, equipment, and lighting. From lighting to thermostat control products, smart home offers a variety of products that homeowners can purchase to begin saving on energy costs today.

2.8.1   Heating and Cooling

As shown in Figure 2.18, nearly half of a typical utility bill goes toward heating and cooling in a typical house. A programmable thermostat offers the flexibility

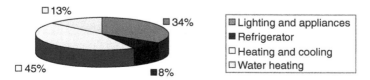

FIGURE 2.18   Percent of residential energy usage.

and power to control the climate in your home efficiently to save energy and lower energy bills. With a programmable thermostat, you can set the temperature to different levels during set times throughout the week. For example, during the winter, you can set the inside temperature to a lower level when you are at work and the house is unoccupied. This can save you nearly $150 on your yearly utility bills depending on climate, home insulation, and other factors. With home automation technology, one can lower the thermostat setting and also turn off all the lights and appliances in your home by hitting a single preprogrammed button on the way out the door.

2.8.2 Lights

While 34% of a typical energy bill goes toward lights and appliances, a full 25% of the utility bill is actually spent on just lights. Reduce your energy usage by simply turning off lights when you do not need them with automatic timers and motion detectors. You can also use light dimmers to reduce wattage and output to save energy.

2.8.3 Automatic Timers

Program timers are used to turn on holiday and/or porch lights after sunset and off at bedtime.

2.8.4 Motion Sensors

Never forget to turn lights off by using motion sensors that automatically turn off lights when a room is left unoccupied for a long period of time – perfect for garages, hallways, and bathrooms. Occupancy sensors can cut lighting costs by as much as 50%.

2.8.5 Light Dimmer

You often do not need the full brightness of lights in a room, especially in the den while watching TV or having a romantic meal in the dining room. Use technology to preset brightness levels to match the occasion.

2.8.6 Energy-Efficient Light Bulbs

It only takes 18 s to change a light bulb. Save money and energy by swapping your existing incandescent bulbs for energy-efficient compact fluorescent lights (CFLs). For the same amount of light, the CFLs use up to 75% less energy and also last 10 times longer, according to Home Energy Saver. Click here to see our fluorescent bulbs and LEDs.

2.9 MANAGING ENERGY SMART HOMES ACCORDING TO ENERGY PRICES

Local energy production and consumption means in a Smart Home can be managed by a building energy management system. Advanced BEMS makes it possible to deploy new kinds of energy management strategies that may change the way of consuming and producing energy by supporting occupants to reach a better energy performance and comfort. A smart home is a residential dwelling equipped with sensors and possibly actuators to collect data and send control according to occupants' activities and expectations [18,20]. Potential applications for smart homes are described in Ref. [21]. The goal of these applications is to improve home comfort, convenience, security, and entertainment. Thanks to the communication network, a load management mechanism has been proposed in Ref. [22]. Since then, several studies have been conducted in order to design an optimized electric BEMS able to determine the best energy assignment plan according to a given criteria. In Ref. [23], an analysis of the load management technique is detailed. According to Ref. [24], energy management system contains methods that coordinate the activities of energy consumers and energy providers in order to best fit energy production capabilities with consumer needs. With such solutions, electricity can be reduced to support the grid. During the last 2 years, many research projects have focused on demand side management and loads control of domestic smart grid technologies for many reasons. First, energy use in buildings currently account for about 32% of total global energy consumption. In terms of primary energy consumption, buildings represent around 40% in most IEA (International Energy Agency) countries [25] and 65% of the total electric consumption [26]. Buildings are also responsible for 36% of the EU CO_2 emissions. Not only energy performance, but also load management in buildings is a key issue to achieve the EU climate and energy objectives, namely, 20% reduction of the greenhouse gases emissions by 2020 and 20% energy savings by 2020 [27]. These technologies may modify the domestic energy use (electricity and heat) and adjust the electricity consumption/production in dwellings [28,29]. These research works can be divided into two complementary categories: predictive energy management and real-time control. This control uses prediction model in addition to measured data in order to forecast the optimum control strategy that will be implemented. Similar researches have been carried out on predictive controllers using stochastic models [30]. Both short-term (10–20 min) and long-term (days) prediction errors lay within acceptable ranges in terms of both temperature and humidity levels. The second category of research also uses the predictive control, but it introduces real-time control algorithms in order to give more benefits contrary to Refs [31,32], which do not study the price prediction. Most of these researchers studied the real-time electricity pricing environments to encourage users to adjust load peaks for two goals: reducing their electricity bill and reducing the peak-to-average ratio (PAR) in load demand [33,34]. The BEMS are usually based on simple models because it is difficult to determine the parameters of detailed models that fits actual measurements. BEMS has to be "appropriate" to detailed models. The problem of the evaluation of the degree of "appropriation" and then the evaluation of the proposed solutions by the BEMS is rarely treated as a research problem. This work deals with an analysis of a global model-based anticipative building energy management system (GMBA-BEMS) managing household energy. Most anticipative approaches of energy management problem focus on specific appliances such as electrical water heater in Ref. [34] and HVAC (heating, ventilation, and air-conditioning) system in Ref. [33]. HVAC is the technology of indoor and vehicular environmental comfort. HVAC system design is a subdiscipline of mechanical engineering, based on the principles of thermodynamics, fluid mechanics, and heat transfer. According to Ref. [35], the approach of GMBA-BEMS called G-home Tech used in this paper is general enough to handle a large set of electric appliances: electrical heater, washing machine, dishwasher, fridge, and so on. It represents 80% of the total residential consumption [36].

As detailed below, BEMS are based on simple models because it is difficult to determine the parameters of detailed models that fit actual measurements; BEMS has to be "appropriate" to details models: It requires validation scenarios and a building simulator connected to a BEMS. Regarding the proposed test bench, the energy management strategy aims to minimize the household's electricity cost taking into account price signals from the grid by optimally scheduling the operation and energy consumption of each appliance according to user comfort expectations. As in Ref. [33], a time-varying curve of electricity price is used. The household load management is based on price and consumption forecasts considering users' comfort to meet an optimization objective that compromises minimum payment and maximum comfort. Real-time adjustments are then done according to real-time electricity market prices actual interest. However, in this study, it is done according to the total available power: the PV production (according to solar radiation curves) and the power limitation (subscription), to the electricity market prices, and the power consumption of the other appliances (Figure 2.18). Note that PV is a method of generating electrical power by converting solar radiation into direct current electricity using semiconductors that exhibit the photovoltaic effect. Photovoltaic power generation employs solar panels composed of a number of solar cells containing a photovoltaic material.

The validation test bench is not only concerned with the heating control such as in Refs [37,38] but also with the electrical appliances making the problem more complex. It aims to introduce a real-time energy management decision based on both reactive and anticipative global algorithms [39] contrary to Refs [40,41] where the authors use a predictive control to anticipate a solution for heating systems [35] and Ref. [42] details the BEMS algorithm selected for

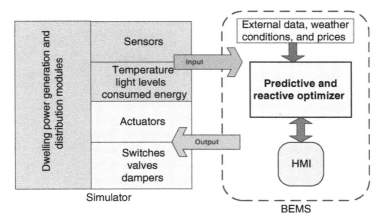

FIGURE 2.19 Virtual cosimulation general scheme for BEMS validation.

the proposed analysis. The anticipative layer assigns energy references by taking into account predicted events. Concerning the reactive layer, it intervenes when the anticipative plan cannot be followed because of unforecasted events and it decides whether some appliances have to be switched *on* or *off*. On the other hand, to validate a BEMS, two parts must be presented: the simulator and the energy management algorithms (Figure 2.19). In the BEMS, the multilayers algorithms are in interaction with external data that come from the weather, the energy marketer, the human–machine interface (HMI), and the real-time simulator. The HMI can be used by the occupant to provide instructions to the BEMS. Simulators replace a dwelling and its HVAC systems to simulate their response to the BEMS as described in Ref. [38]. They are used to improve product development, to train BEMS operators, to tune actuators, and to simulate faulty situations [43]. Then, the validation of a BEMS should be done through a simulator model. The simulation models include in addition to the HVAC many electrical appliances such as lighting, flaps, washing machine, dishwasher, and fridge. Some simulators are presented in the literature. For example, the software "PME Comfort" is used to simulate the thermal comfort of dwelling [44]. "Solene" [45] simulates the sunshine, light, and radiation. "ESP-r" [46] and "FLOVENT" [47] simulate the movement of air in dwellings.

The BEMS is fed up with simplified models compatible with a mixed integer linear programming formulation (Figure 2.20) [48]. But the behavior of a real dwelling is much more complex. Therefore, simulation requires finer dwelling models than the BEMS. To summarize, the objective of the chapter is to analyze a BEMS in a context of variable pricing. To do this, three steps must be performed: choose building simulator, configure the GMBA-BEMS, and analyze the results.

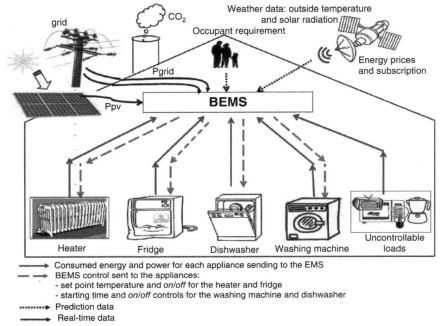

- ──────▶ Consumed energy and power for each appliance sending to the EMS
- ── ──▶ BEMS control sent to the appliances:
  - set point temperature and *on/off* for the heater and fridge
  - starting time and *on/off* controls for the washing machine and dishwasher
- ·········▶ Prediction data
- ──────▶ Real-time data

FIGURE 2.20   Exchanged command from BEMS and data from appliances.

2.10   SMART ENERGY MONITORING SYSTEMS TO HELP IN CONTROLLING ELECTRICITY BILL

A connected smart home energy monitoring system makes it easy to view your electricity usage and save money. We have written extensively on the benefits of home automation for energy management, particularly on how smart thermostats that allow you to easily adjust your home's temperature, even while you are away, to save money and make your house more comfortable. However, there is one important aspect that we left out – smart energy management is more than just automating how your thermostat goes up or down. Real energy management requires an energy monitoring system, which means knowing how much energy you are using. If the amount and time of energy of a certain home are known, one can better respond to that usage and take control of that home energy costs. Before smart home systems, energy monitoring mostly meant scanning your electricity bill each month and then telling your family to shut off the lights. New technology makes the process much easier.

There are several ways you can monitor (and then respond to) your home's electricity usage. Some methods let you monitor only the appliance you have connected to the monitoring device, while other systems take a whole-house approach. Here are a few energy monitoring systems worth checking out.

The Home Energy Monitoring System uses your home's existing power lines to monitor energy usage in real time. It can actually tackle up to 32 individual circuits, as well as individual rooms. That way, you can keep tabs on devices that are energy hogs and unplug them accordingly.

The electricity monitor uses measuring devices that clamp onto the main conductors inside your breaker panel. The devices then send the data that are measured over your home's power lines. No extra wiring is needed! Those data are collected in a receiving unit, which you can plug into any outlet around the house. Inside, the receiving unit has the company's Footprints software, which can store and track up to 10 years' worth of energy data. Homeowners can keep tabs on that information via any web-enabled device (e.g., smartphone). You can even opt to receive customized alerts via text or e-mail messages. Other features include colored LEDs to alert users to different parameters, as well as the option to turn loads on and off based on the cost of electricity and use.

2.11 ADVANCING BUILDING ENERGY MANAGEMENT SYSTEM TO ENABLE SMART GRID INTEROPERATION

The smart grid is a nationwide project to modernize the 100-year-old power infrastructure by integrating the state-of-the-art information technology for two ultimate goals: (i) balance the power demand (consumption) with the supply via active interoperation among energy resources and (ii) accelerate the use of environment-friendly renewable energy sources. In the smart grid context, building facilities, including industrial, commercial, and residential sectors, have been primary energy consumers; they consume 72% of total energy in the United States [49]. In the future, the facilities will be capable of generating and storing energy with the potential inclusion of electric vehicles (EVs), solar panels, and batteries. To manage such complicating energy resources, the research community has developed an intelligent energy management system. Its eventual goal is to maximize energy efficiency in a building and to minimize the electricity cost by *making the best use of energy resources available in a building*. To this end, the EMS communicates with individual building equipment to collect its energy data and to control them separately: fine-grained management. The EMS, then, analyzes the data collection so as to detect any inefficient building operations and failures.

From the smart grid perspective, the customer's building facility becomes the most important entity to interoperate, as its energy capability (demand, generation, and storage) dramatically increases. For instance, when the bulk power source confronts a shortage of power supply, the customer is able to reduce current power consumption, which can prevent blackouts. The EMS in the facility is required to support such *smart grid interoperation* by enabling customer energy resources to interact with other systems outside the facility. However, the existing EMS has been designed as a stand-alone system without any consideration of the interoperation aspect.

To resolve the issue, we propose a new design of EMS, named *premises automation system* (PAS). PAS aims at accommodating both the customer need of efficient energy management and the grid need of customer interoperation.

To address the customer need, PAS inherits fundamental design issues from the existing EMS model. It connects to customer energy resources that use heterogeneous communication protocols and technologies and manages them in a fine-grained manner. To address the grid need, we first review existing and potential energy services that realize the customer interoperation and then classify them into two categories: grid services and customer services. Under grid services, the customer facility receives and consumes service data delivered from the smart grid, while the facility provides service data to the grid in a customer service. For each category of services, we examine functional requirements of the EMS in the four aspects of service data type: a communication interface to realize the service, required intelligence (data processing and knowledge generation), security, and privacy.

To demonstrate the feasibility of PAS, we develop and deploy a testbed in our campus. In the testbed, PAS connects to and manages various types of energy resources, consumes an automated demand response service, generates valuable energy forecast data, and provides energy services to smart grid based on a service model in a secure manner.

2.11.1 Smart Grid and Customer Interoperation

The smart grid aims at making the existing power grid more *intelligent* and *interoperable* by allowing bidirectional flows of information and electrical power. By integrating the state-of-the-art information and communication technologies to the power infrastructure, energy resources with embedded sensors generate valuable data that are then shared with all other resources in the smart grid. Such information flow enables smart grid to monitor the status of power generation and consumption accurately and respond quickly to potential failure. The smart grid also allows a bidirectional flow of electricity, compared with today's power grid electricity flows from central bulk generators to end consumers. Various types of renewables can be installed on the consumers' side and supply power back to the grid reversely. On top of the information and power network, the operational goal of a smart grid system is to maximize interoperations among energy resources so as to balance the power demand with the supply, which eventually makes the power grid more reliable and sustainable. To facilitate the interoperations, National Institute of Standards and Technology (NIST) presents a conceptual model consisting of seven domains, each of which represents a high-level grouping of smart grid entities having similar objectives [50].

Customer domain in the conceptual model represents customer facilities (e.g., office, campus, and home) that consume more than 70% of total energy in the United States. Traditionally, a building automation system (BAS) controls facility equipment for the purpose of occupant comfort and optimal business operations.

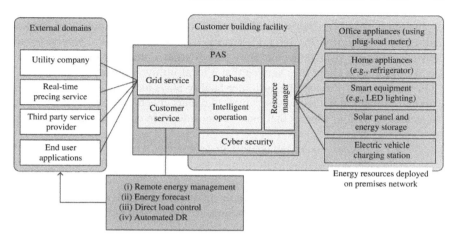

FIGURE 2.21   An information system architecture around the customer domain.

Today, the introduction of smart grid changes the customers' awareness and expectation about their energy management. They want to see breakdowns of energy usage and to take actions to reduce energy costs. Moreover, they are interested in instrumenting new types of energy resources like solar panel within the facilities.

To meet the emerging customer needs, recent research on the customer domain has developed an advanced energy management system to build a smart building. It performs fine-grained energy measurements and controls, say, at individual home/office appliance level. It optionally analyzes the collected data and controls equipment in a way to maximize the energy efficiency inside the customer facility. Although the existing EMS research makes the customer facility more intelligent to satisfy the *customer needs*, it has barely taken the *grid need of interoperation* into consideration.

As the customers' capabilities of energy consumption, generation, and storage increase, it becomes most important to interoperate with the facility for the purpose of energy balance in the smart grid. And the EMS is expected to play a gateway role interconnecting the facility to other domains for the interoperation. Thus, the design of the EMS must be enhanced so as to enable customer energy resources to interact with other smart grid entities outside the facility, that is, supporting customer interoperation. Figure 2.21 shows information system architecture around the customer domain, including PAS, a premises network, energy resources, external domains, service providers, and energy services.

2.11.2   Customer Interoperation and Energy Service

Customer interoperation is an interaction of the customer facility with external domains in which the customer's own resources are engaged.

The interoperable energy services are divided into two categories: grid service and customer service. In the grid service, a customer facility receives and consumes service data delivered from an external domain. For instance, the facility becomes a client of a service that a local utility company provides.

In the customer service, the customer facility plays as a service provider, and external domains use facility's services as clients. Each category of service is characterized by four aspects that must be considered in customer interoperation:

(i) *Service data* could be energy measurement, energy forecast, control message, conventional information such as weather forecast, and power price.

(ii) *Service interface* enables interdomain communications of which there are three issues of interface abstraction, data representation, and interaction model.

(iii) *The intelligent unit* performs interpretation of external data, knowledge generation, and decision making to take energy-related actions.

(iv) *Security* addresses the most critical security concern at each category of energy service.

Taking these aspects into consideration, the smart grid energy management system can be designed, and sometimes called *premises automation system*.

2.12   COMMUNICATION FOR BEMS

Building automation systems provide automatic control of the conditions of indoor environments. The automation of heating, ventilation, and air-conditioning systems in large functional buildings is the historical root and still core domain of BAS. The primary goal is to realize significant savings in energy and reduce cost. Yet, the reach of BAS has extended to include information from all kinds of building systems, working toward the goal of "intelligent buildings." Since these systems are diverse by tradition, integration issues are of particular importance. When compared with the field of industrial automation, building automation exhibits specific, differing characteristics. This chapter introduces the task of building automation and the systems and communications infrastructure necessary to address it. Basic requirements are covered as well as standard application models and typical services. An overview of relevant standards is given, including BACnet, LonWorks, and EIB/KNX as open systems of key significance in the building automation domain. This chapter focuses on the automation of large functional buildings, which in the following will be referred to as "buildings" for simplicity. Examples include office buildings, hospitals, warehouses, or department stores as well as largely distributed complexes of smaller installations such as retail chains or gas stations. These types of buildings are especially interesting since their size, scale, and complexity hold considerable potential for optimization, but also challenges.

2.12.1   Building Automation System

Building automation is the automatic centralized control of a building's heating, ventilation, and air-conditioning, lighting, and other systems through a building management system or building automation system. The objectives of building automation are improved occupant comfort, efficient operation of building systems, reduction in energy consumption and operating costs, and improved life cycle of utilities.

Building automation is an example of a distributed control system – the computer networking of electronic devices designed to monitor and control the mechanical, security, fire and flood safety, lighting (especially emergency lighting), HVAC and humidity control, and ventilation systems in a building (Figure 2.22).

BAS core functionality keeps building climate within a specified range, provides light to rooms based on an occupancy schedule (in the absence of overt switches to the contrary), monitors performance and device failures in all systems, and provides malfunction alarms to building maintenance staff. A BAS should reduce building energy and maintenance costs compared to a noncontrolled building. Most commercial, institutional, and industrial buildings built after

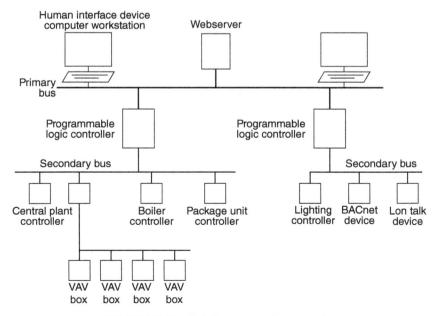

FIGURE 2.22   Building automation example.

2000 include a BAS. Many older buildings have been retrofitted with a new BAS, typically financed through energy and insurance savings, and other savings associated with pre-emptive maintenance and fault detection.

A building controlled by a BAS is often referred to as an intelligent building, "smart building," or (if a residence) a "smart home." Commercial and industrial buildings have historically relied on robust proven protocols (like BACnet), while proprietary protocols like X-10 were used in homes. Recent IEEE standards (notably IEEE 802.15.4, IEEE 1901 and IEEE 1905.1, IEEE 802.21, IEEE 802.11ac, IEEE 802.3at) and consortia efforts like nVoy (which verifies IEEE 1905.1 compliance) or QIVICON have provided a standards-based foundation for heterogeneous networking of many devices on many physical networks for diverse purposes, and quality of service and failover guarantees appropriate to support human health and safety. Accordingly, commercial, industrial, military, and other institutional users now use systems that differ from home systems mostly in scale.

Almost all multistory green buildings are designed to accommodate a BAS for the energy, air, and water conservation characteristics. Electrical device demand response is a typical function of a BAS, as it is the more sophisticated ventilation and humidity monitoring required of "tight" insulated buildings. Most green buildings also use as many low-power DC devices as possible, typically integrated with power over Ethernet wiring, so by definition always accessible to a BAS through the Ethernet connectivity. Even a Passivhaus design intended to consume no net energy whatsoever will typically require a BAS to manage heat capture, shading and venting, and scheduling device use.

2.12.2 Busses and Protocols

Most building automation networks consist of a *primary* and *secondary* bus that connect high-level controllers (generally specialized for building automation, but may be generic programmable logic controllers) with lower level controllers, input/output devices, and a user interface (also known as a human interface device). ASHRAE's open protocol BACnet or the open protocol LonTalk specify how most such devices interoperate. Modern systems use SNMP to track events, building on decades of history with SNMP-based protocols in the computer networking world.

Physical connectivity between devices was historically provided by dedicated optical fiber, Ethernet, ARCNET, RS-232, RS-485, or a low-bandwidth special-purpose wireless network. Modern systems rely on standards-based multiprotocol heterogeneous networking such as that specified in the IEEE 1905.1 standard and verified by the nVoy auditing mark. These accommodate typically only IP-based networking, but can make use of any existing wiring, and also integrate power line networking over AC circuits, power over Ethernet low-power DC circuits, high-bandwidth wireless networks such as LTE and IEEE 802.11n and IEEE 802.11ac and often integrate these using the building-specific wireless mesh open standard ZigBee).

Proprietary hardware dominates the controller market. Each company has controllers for specific applications. Some are designed with limited controls and no interoperability, such as simple packaged roof top units for HVAC. Software will typically not integrate well with packages from other vendors. Cooperation is at the Zigbee/BACnet/LonTalk level only.

Current systems provide interoperability at the application level, allowing users to mix-and-match devices from different manufacturers and to provide integration with other compatible building control systems. These typically rely on SNMP, long used for this same purpose to integrate diverse computer networking devices into one coherent network.

The communication protocol BACnet was specially developed for the requirements of buildings. It is suited for both the automation and the management level. The emphasis is placed on building automation and control with a view to HVAC plants, fire control panels, intrusion detection, and access control systems. BACnet is continually being extended for additional building-specific systems such as escalators and elevators. By integrating new IT technologies such as IPv6 and Web services, the BACnet standard is further developing into a modern, IT-friendly, and multidisciplinary building protocol. At the same time, standardized ASHRAE or AMEV device profiles ensure a high level of quality and planning reliability with a strict testing and certification procedure.

- Highest investment protection, thanks to the use of the open, worldwide ISO 16484-5 standard.
- Continued incremental development by ASHRAE, always focusing on the requirements in and around buildings.
- Vendor-independent.
- No license fees.
- Guaranteed reliability, thanks to independent test houses and certification bodies for BACnet devices.
- Different transmission media, such as BACnet IP, BACnet LonTalk, or BACnet MS/TP can be combined and support the most flexible topologies.
- Integration of the most diverse types of plants and vendors without having to use special hardware.
- Siemens is involved in the BACnet organizations worldwide in order to promote the development of the standard.

KNX is an open, worldwide standard used for more than 20 years, conforming to EN 50090 and ISO/IEC 14543, which is supported by more than 300 vendors. With KNX technology, advanced multiple disciplines as well as simple solutions can be implemented to satisfy individual requirements in room and building automation in a flexible way. KNX products for the control of lighting systems, shading, and room climate plus energy management and security functions excel in ease of installation and commissioning. A vendor-independent tool (ETS) is available for commissioning. KNX can use twisted pair cables, radio frequency (RF), or data transmission networks in connection with the Internet Protocol for communication between the devices. Coordinated room and building management often demands the integration of other technologies and systems. Hence, KNX links and interfaces for connection to Ethernet/IP, RF, lighting control with DALI, and building automation and control systems are provided.

- Investment protection and interoperability, thanks to the standardized, worldwide KNX standard.
- Highest level of comfort and security while ensuring low energy consumption.
- Matching products and systems for comprehensive room and building automation.
- Straightforward connection to higher level building automation systems.
- Vendor- and product-independent commissioning software provides standardized commissioning procedures (ETS).
- Choice of transmission media: KNX TP, KNX RF, and KNX IP.
- Corresponds to the former European Installation Bus (EIB) and is backward-compatible.
- Siemens is member of the KNX Association and actively takes part in the evolution of the KNX standard.

KNX PL-Link fully complies with the KNX standard. Communication between the room automation stations PXC3 of Desigo Total Room Automation (TRA) and peripheral devices with KNX PL-Link has been optimized within the framework of the KNX standard to the extent that plug-and-play functionality is available with automatic device recognition. KNX PL-Link devices are configured using the Desigo tools. The KNX commissioning software (ETS) is not needed.

- Automatic recognition of devices with KNX PL-Link
- Simplest configuration of devices with KNX PL-Link using the Desigo tools.
- Comprehensive portfolio of devices with KNX PL-Link for all technical disciplines in the room.
- Integrated monitoring of devices with KNX PL-Link using room automation stations PXC3.
- Replacement of a device with KNX PL-Link without any tools.
- Two-wire standard cable for up to 64 peripheral devices in line or star topologies with a maximum line length of 1000 m.
- Feeding of up to 64 peripheral devices directly via the bus line.
- Fast event-oriented communication for lighting and shading applications.

- Room automation stations PXC3 allow simultaneous integration of devices with KNX PL-Link and KNX S-Mode on a single bus line.
- Devices with KNX S-Mode are commissioned using ETS.

The LonWorks-based communication protocol is one of the most widely deployed technologies worldwide. Using the protocol, complete networks made up of interoperable products can be created. This is proven by the fact that more than 700 LonMark®-certified products from more than 400 companies in the fields of building automation and control, traffic, and energy supply are used. Owing to its worldwide use and being a global standard, LonWorks is also of great importance to Siemens, focusing on HVAC functions in room automation and at the field level.

The protocol conforms to ISO/IEC 14908 (worldwide), EN 14908 (Europe), ANSI/CEA-709/852 (the United States), and is also standardized in China.

- LonWorks is suited for use with different types of transmission media, such as twisted pair cables, power line, RF, fiber optics, or IP (both TCP/IP and UDP/IP), which makes it very flexible.
- Straightforward installation with a choice of different cabling topologies (e.g., star or line).

    The connection of objects via bindings (e.g., standard network variables (SNVTs), standard configuration properties (SCPTs)) can be defined at the project engineering stage or can be adapted in the field. This simplifies the engineering process and helps prevent errors.
- Siemens is involved in the organization LonMark® International with the objective to protect and further develop the standard.

Siemens Products Featuring LonWorks Communication

- Desigo RXC room controllers
- Room operator units QAX5x.x
- Climatix lines

DALI (Digital Addressable Lighting Interface) is a standardized interface for lighting control. Electronic ballasts for fluorescent lamps, transformers, and sensors of lighting systems communicate with the building automation and control system via DALI.

- Extensive installation capacity and system flexibility, thanks to the support of up to 64 electronic ballasts, 16 groups, and 16 scenes.
- Increased reliability owing to bidirectional communication with feedback of operating state (dim level, lamp failure, etc.).

- Polarity-free two-wire link in line, star or mixed topology with a maximum cable length of 300 m.
- Individually addressable operating units with free and flexible assignment of lamps with no need for making wiring changes.
- Integration of emergency lighting in general lighting systems.
- Siemens is a member of the work group DALI and participates, therefore, actively in the further development of the standard.

Worldwide leading companies operating in the field of building infrastructure have joined to form the EnOcean Alliance, aimed at implementing innovative RF solutions for sustainable building projects. Core technology is the self-powered RF technology developed by EnOcean for maintenance-free sensors, which can be installed wherever desired. The EnOcean Alliance stands for the incremental development of the interoperable standard and for a secure future of the innovative RF sensor technology.

- EnOcean combines wireless communication with methods developed to produce energy, aimed at minimizing product maintenance and the number of batteries in use.
- Standardized EnOcean communication affords access to a large number of easy-to-integrate field devices.

Modbus is an open and widely used de facto standard applied in a large number of application areas, such as the industrial sector, buildings, traffic, and energy. The Modbus protocol is used to establish masterslave/client-server communication between intelligent devices. Using Modbus, a master (e.g., automation station) and several slaves (e.g., chillers) can be interconnected. Data transmission takes place through one of the three operating modes: Modbus ASCII, RTU, or TCP.

M-Bus (Meter Bus) www.m-bus.com M-bus is a European standard covering remote readout of meters and can be used with different types of consumption meters and various types of valves and actuators. Data (e.g., heat energy) can be read out electronically. In that case, transmission is serial via a two-wire line with reversed polarity protection, from the connected slaves (meters) to a master. M-bus meters are available for the acquisition of heat, water, electricity, and gas.

OPC www.opcfoundation.org OPC is a standardized software interface facilitating the exchange of data between different types of devices, control systems, and applications of different vendors. This interface is frequently used to collect the process values of third-party devices for further handling by a building automation and control system.

Web (IT Standard Technology) This is a generic term for a number of standardized communication protocols used in the IT world, be it within a local plant or via the Internet. Included are protocols that users work with when communicating with plants and products, such as graphic user interfaces that can be operated via Web browsers and/or touch panels, e-mail messages to service personnel, or implementation of firmware changes. In addition, this comprises an increasing number of protocols for direct communication between machines, such as the exchange of device management information, or so-called Web services for the connection of plants, even beyond the boundaries of building automation (e.g., to external building or energy management systems).

Standardized Communication Protocols for More Economical Operation
Open communication in building technology is important and facilitates the straightforward and secure integration of third-party systems at all levels. In the field of building automation, all communication protocols listed above are supported with no restriction to standards as in Figure 2.23. These are communication standards that were developed for the successful creation and maintenance of projects. These make communication more secure, support efficient engineering, and simplify maintenance and interoperability, thereby improving the investment protection.

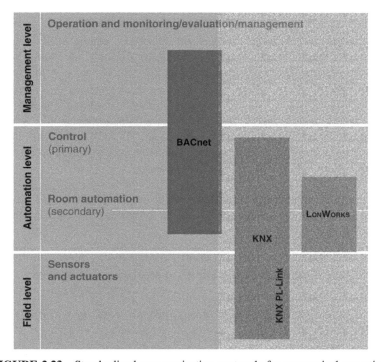

FIGURE 2.23   Standardized communication protocols for economical operation.

The Building Technologies Division of Siemens supplies complete building automation and control systems and – in addition to heating, ventilation, and air-conditioning – integrates lighting, shading, fire safety and security, lifts, distribution of electrical energy and other forms of energy, and so on.

Building automation and control systems from Siemens and solutions based on them make use of the standards listed above. The standardized and independent communication protocols are subject to incremental development and ensure a consistent exchange of information between devices and systems.

2.13 DATA MANAGEMENT FOR BUILDING

To allow for a better structuring of the diversity of technical systems, it is worth taking a closer look at the various building management tasks. Usually, the following three task areas are distinguished:

- *Commercial management* is performed by specialized systems that support the company's business processes and comprises various subareas from purchasing to logistics to sales and maintenance. These systems are more or less integrated, depending on the solution, and can be combined under the name ERP (Enterprise Resource Planning). Among the most well-known companies in this field are SAP and Oracle, for example.
- *Infrastructural building management* comprises, among other things, systems for the maintenance of the building, for example, the facility management systems (FMS), which manage the maintenance of the technical facilities.
- *Technical building management* comprises the building automation and security management. While the building automation deals with, for example, heating, ventilation, air-conditioning (HVAC), light, and lifts, the security management deals with fire detection, burglar alarm, access control, video surveillance, and other security topics.

The operating behavior of the systems can be optimized in a simple manner via the management station and provides for an energy-efficient operation of the entire building installations.

2.13.1 Main Functions of the Building Management System

- Operator control and monitoring
  Fast and selective monitoring and operation of the system with practical plant and room diagrams.
- Time programs
  Central programming of all time-controlled building functions.

- Alarm handling

  Detailed overview of the alarms for a fast localization and elimination of faults. Central elements of the alarm handling are therefore the danger identification, danger alarm, and an adequate intervention. This is supported by the flexible transmission of alarms to mobile devices, for example, printers or pagers.

- Event control

  System-wide monitoring of systems and processes with regard to the occurrence of certain criteria for the triggering of certain predefined actions.

- Reporting

Modern management stations today work with integrated database applications. This allows for the storage of an almost unlimited number of past events and their recorded handling. With these plant-specific records and the corresponding query options, the following questions can be answered, for example:

- What has happened in the past 24 h?
- How many interferences occurred within a certain period?
- Who has done what and when following yesterday's burglar alarm?

To provide this main function, a whole range of additional functions is required, which so to speak forms the infrastructure of the building management system. The most important of these additional functions are, for example, access rights concept, user administration, password administration, object management in tree and graphic structures, and graphic level management.

2.13.2   Planning of a Building Management System

Within the scope of BMS projects, there are individual project phases comprising different contents and responsibilities. The first project phases are described in the following.

The customer/user must define objectives; this serves as a basis for the preparation of a requirements specification.

The definition of objectives comprises the following elements:

- Scope of the building automation and security management subsystems to be integrated.
- Definition of the integration:

Combining all subsystems of building installations (fire alarm, gas warning, burglar alarm, access control or video surveillance, HVAC systems, lighting, and further external systems) by integrating them into a building management system brings the following advantages:

- Improved overview and thus increased safety.
- Lower costs in comparison to several independent control centers with regard to acquisition, configuration, and maintenance.
- Consistent operational concept and thus less training time and effort, and no danger of confusion in an emergency.
- Only one system has to be integrated into the in-house IT infrastructure.
- Interactions between the subsystems are possible in a much easier way.
- Alarm escalation and alarm transmission is done in a more uniform way.
- Integrated video systems allow for a direct view of the fault cause.

The objective is, therefore, to integrate all subsystems as completely as possible into the building management system.

- Expected improvements compared to a single-system solution.
- Demands on the failure safety (redundancy solutions).
- Demands on the power supply (e.g., UPS).
- Description of the workplaces and tasks of the employees at the workplaces.

From the definition of objectives, a requirements specification has to be prepared with the collaboration of the user and planner.

2.14 POWER MANAGEMENT

Due to the increasing energy costs, saving energy becomes more and more important on every sector. At the same time, ecological goals are to be attained, for example, the specifications with regard to the reduction of emissions and greenhouse gases. This results first of all in the selection of energy-efficient components, but it also necessitates an ecological and economic power management.

In the planning phase, the property costs are to be kept as low as possible, while the operator is interested in minimizing the operating costs. When planning the electrical power distribution, the basics for the power management should be established. The following aspects are to be taken into account:

- Provide the required components with interfaces for measurements and sensors.
- Use standardized bus systems and communication capable devices.
- Ensure expandability (e.g., expandable cable laying and installation of transformers in cabinets) to keep interruptions during operation at a minimum.

The focus of a power management system is on the request for improved transparency of energy consumption and energy quality as well as on ensuring the

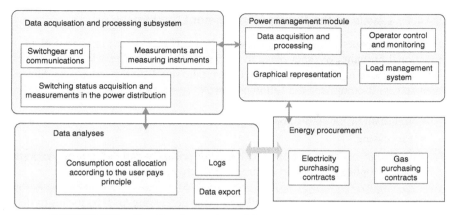

FIGURE 2.24   Functional overview of the power management system.

availability of power distribution. An all-round transparency is the basis for an optimization of energy costs and consumption. The obtained information provides a realistic basis for a cost center allocation as well as for measures to improve the energy performance. Moreover, savings are documented. The functional overview of the power management system is illustrated in Figure 2.24.

Functions of the power management system are summarized as follows:

- Analysis of the energy data/energy flows with specific load curve diagrams.
- Visualization of the interdependencies.
- Detection of savings potentials, assessed minimum and maximum values.
- Energy measurements for accounting purposes (internal cost center allocation, external billing).
- Benchmarking, internal (product line/building part) or external (property/ installations with comparable use based on obtained measured values).
- Visualization of the power supply with switching states and energy flows.
- Preparation of decisions, for example, for power supply extensions.
- Verifiable efficiency improvements.
- Targeted troubleshooting via fast and detailed information on events and faults that occur in the power distribution within the installations/ building.
- Logging of fault and event messages (e.g., switching sequences) with a date and time stamp so that downtimes can be documented and fault processes can be traced and analyzed later using the data recorded.
- Compliance with purchasing contracts via the selective control of consuming devices.
- Automatic notification of the service personnel.

2.14.1 Levels of the Power Management System

Power management is the special energy point of view of an industrial plant, a functional building, or other piece of property. The view begins with the energy import, expands to its distribution, and ends at the supply to the consuming devices themselves. It comprises the following levels:

- Acquisition for status and measurements.
- Processing.
- Operator control and monitoring with visualization, archiving, reports, import optimization, and control of switchgear.

The data acquisition level is connected to the processing level by means of field buses and the processing level communicates with the visualization system and data archive via LAN (Local Area Network) as illustrated in Figure 2.25.

The acquired status information is depicted on the status displays in the control center, thus enabling remote control. Measured value readings are displayed.

2.14.2 Switching Status Acquisition and Measurements in the Power Distribution

In order to command of optimum purchase/consumption quantity records during the utilization phase, the required measuring points and the power distribution components to be monitored must be planned and configured at an early stage.

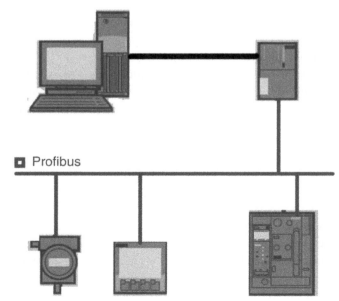

■ Profibus

FIGURE 2.25 Profibus connects the acquisition and processing level.

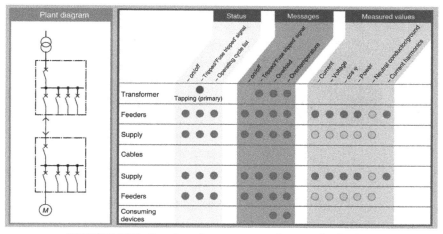

FIGURE 2.26 Levels and components of power distribution.

Important information for that

- Types of energy
- Components of the power supply (e.g., also UPS, emergency generators)
- Division of the power demand according to the planned scenarios of use

For the various levels and components of power distribution (Figure 2.26), it has to be taken into account which measurements and messages are required during operation as well as the various requirements for:

- Critical areas/consuming devices (availability)
- Billing values (plausibility, contract monitoring, cost center management)
- Transparency for operation (measured values, status)
- Utilization (expansions, energy import monitoring)

2.14.3 Switchgear and Communications

The basis of each power management system are the measured values and data from the field level in which the energy in consumed. A large number of devices can already be evaluated via bus systems such as Profibus by a power management system with regard to some specific data.

Circuit-Breaker-Protected Switchgear: Circuit-Breakers Circuit-breaker-protected switchgear can be equipped or retrofitted with the following signals (Figure 2.27):

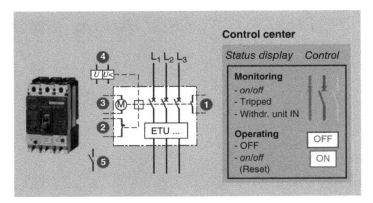

FIGURE 2.27   Circuit-breaker-protected switchgear.

1. The auxiliary *on/off* switch signalizes the status of the circuit-breaker, *on* or *off*.
2. The alarm switch signalizes whether the breaker has tripped.
3. The motorized drive acts on the switching rods and permits remote control of the breaker.
4. The release operates in parallel to the overcurrent release and acts directly upon the switch-off mechanism of the circuit-breaker. Voltage and undervoltage releases are to be distinguished as follows: voltage releases switch when voltage is applied, undervoltage releases switch when voltage is interrupted.
5. The alarm switch signalizes the status of the withdrawable unit. Only if all withdrawable circuit-breaker units have been properly pushed in (i.e., contacted) can electric energy be switched.

Control Center   The visualization screen shows the circuit-breaker status with the aid of the pictograph "*on/off*/tripped/withdrawable unit pushed in" and additionally by means of the color coding for "event/fault/acknowledged/not acknowledged."

The circuit-breaker can be operated remotely from the user interface.

Fuse-Protected Switchgear: Switch Disconnector   Fuse-protected switchgear can be equipped or retrofitted with the following signals as shown in Figure 2.28:

6. The auxiliary *on/off* switch signalizes the status of the switch disconnector, *on* or *off*.
7. The fuse monitor signalizes a triggered/tripped fuse.

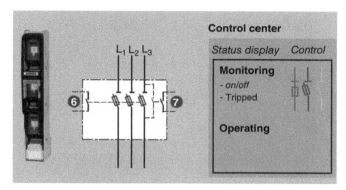

FIGURE 2.28   Fuse-protected switchgear.

Control Center   The visualization screen shows the switch disconnector status with the aid of the pictograph "*on/off*/tripped" and additionally by means of the color coding for "event/fault/acknowledged/not acknowledged." A switch disconnector cannot be operated remotely.

Measurements   Measuring instruments (multifunction instruments, electricity meters, motor management) can produce calculated data (phase displacement, work, power) in addition to current and voltage readings (Figure 2.29).

1. Current transformers convert/transform current measurements into standard values (1 A or 5 A), as the currents typically used in low-voltage distribution (up to 6300 A) cannot be processed directly.
2. The voltage tap directly acquires the voltages applied/measured.

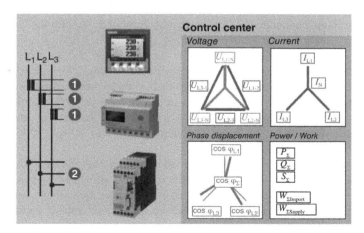

FIGURE 2.29   Measurement procedures.

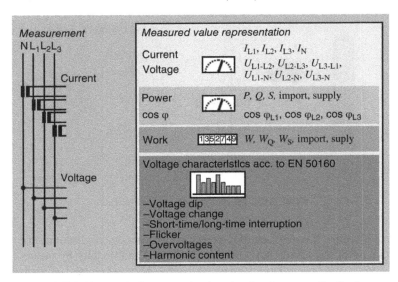

FIGURE 2.30 Typical measured values in electric power distributions.

Control Center The visualization screen shows measurement data for "phase currents/phase voltages/phase displacement/power/work" and also identifies "limit value violations/acknowledged/not acknowledged" by means of the color coding.

Measuring Instruments Measuring instruments acquire current and voltage values in the electric power distribution and, according to their specified scope of performance, they perform the following calculations (Figure 2.30): wattages, phase displacement, work, and voltage characteristics in line with DIN EN 50160 (voltage characteristics of electricity supplied by public distribution networks).

Multifunction Measuring Instruments Built-in device for electric power supply systems with direct measurement display large back-lit high-resolution graphic display, suitable for connection in three-phase networks, in three-wire and four-wire design, for identical loads or different loads, also suitable for single-phase networks, for industrial networks up to 3~690/400 V (e.g., SENTRON PAC4200).

Parameterization can easily be performed by using either the front keys on the instrument panel or the PC-based parameterization software. The number of measuring screens and their contents, that is, measured quantities, can be configured by the user as desired. The instrument has parameterizable digital inputs/ outputs for counter/energy pulses, status monitoring, limit value violations, measuring period synchronization, high-rate/low-rate changeover, and switching to remote control via system software. The measured quantities are summarized as follows:

- Rms values of phase currents and voltages, PEN conductor current.
- Network frequency.
- Active, reactive, and apparent power per phase and for the entire system.
- Electricity meter for high-rate and low-rate price.
- Power factor per phase and for the entire system.
- Symmetry factor of currents and voltages.
- Harmonic contents of voltages and currents.
- Total harmonic distortion (THD).

Electricity Meters

- E-meters for single-phase operation; E-meters for three/four-wire connection
  - Drum-type register for electricity consumed (kWh).
  - S0 interface (pulses).
- E-meter for three/four-wire connection; multirate meter
  - Drum-type registers for electricity consumed (kWh) for high-rate and low-rate price
  - S0 interface (pulses) per rate type
- Multimeter

Built-in modular device for electric power distribution systems with direct measurement display large back-lit graphic display, suitable for connection in three-phase networks, in three-wire and four-wire design, for identical loads or different loads, also suitable for single-phase networks. The measured quantities in this system are summarized as follows:

- Rms values of phase currents and voltages.
- Network frequency.
- Active power per phase and for the entire system.
- Apparent power per phase.
- Reactive power for the entire system.
- Power factor for the entire system.
- Active energy import, export for the entire system at high-rate and low-rate price.
- Reactive energy, inductive and capacitive, for the entire system at high-rate and low-rate price.
- Apparent energy for the entire system at high-rate and low-rate price.

If the meters are to be used for accounting energy quantities, meters that are suitable for an accurate recording of consumptions are required (meters have to be replaced/calibrated at regular intervals). These meters must be identified separately.

Motor Management System   Motor management systems carry out all motor protection and control functions, collect operational, diagnostic and statistic data, and handle the communication between the automation system and the motor feeder. They are parameterized using PC-based parameterization software.

Measured Quantities

- Rms and maximum values of phase currents.
- R.m.s values of phase voltages.
- Active and apparent power for the entire system.
- Power factor for the entire system.
- Phase asymmetry.

Circuit-Breakers   The circuit-breaker (ACB) has a back-lit graphic display for direct value displaying. This display is located at the release, integrated in the circuit-breaker. It can be easily parameterized using a PC-based parameterization software. The number of measuring screens and their contents, that is, measured quantities, can be configured by the user as desired.

Measured Quantities

- Rms values of phase currents, phase voltages, and PEN conductor current.
- Ground-fault current.
- Network frequency.
- Active, reactive, and apparent power per phase and for the entire system.
- Power factor per phase and for the entire system.
- Symmetry factor of currents and voltages.
- Harmonic contents of voltages and currents up to the 29th order.
- Total harmonic distortion (THD).
- Active, reactive, and apparent work for the entire system and their direction.

Other types of energy can be measured additionally using standard interfaces. The following standard interfaces are customary:

- Analog values 0–20 mA.
- Analog values 4–20 mA.
- Analog values ↔}10 V.

- Analog values PT100 for temperatures.
- Pulses for energy quantities.
- Measured values via bus interfaces.

A device-specific block library allows for a direct view of the multifunctional measuring instruments and the device status with a simple integration via Profibus communication. These are called device drivers for multifunction measuring instruments. Blocks are available for faceplates (view) as a user interface for operator control and monitoring that allow for different views to display measured values and to reset limit values for warnings and alarms. Driver block interface to the faceplates and diagnostic blocks are necessary for multifunction measuring instruments. Further device drivers are available as addons for control systems, for example, for SIMOCODE for motor control.

2.14.4   Power Management Module

A power management module, as add-on for control systems, provides blocks for the acquisition, preparation, and representation of energy data and offers special functions up to energy-specific reports. The use of certified blocks and standard interfaces as well as means of the control system provides an integrated application requiring low maintenance effort that is suited for long-term use.

Data Acquisition and Processing

- Complete recording and standardization of energy data from different media as pulses, metered values (work values), or power values.
- Time synchronization or with ripple control signal.
- Buffering of the mean energy and power values.
- Calculation and archiving of the mean power and work values based on a freely definable period in the archive of the control system.
- Determination of the consumption trend for a period based on the current value.
- Open interfaces for customer-specific calculation functions (e.g., amount of heat).
- Block for batch-related energy detection.

ABBREVIATIONS

AHEM	automated home energy management
BAS	building automation system
CFLs	compact fluorescent lights
CoC	code of conduct

EMS	energy management system
EV	electric vehicle
GMBA-BEMS	global model-based anticipative building energy management system
GUI	graphical user interface
HAN	home area network
HMI	human–machine interface
HVAC	heating, ventilation, and air-conditioning system
IEA	International Energy Agency
IoT	Internet of Things
NILM	nonintrusive load monitoring
NIST	National Institute of Standards and Technology
OSGi	Open Service Gateway initiative
PAR	peak-to-average ratio
PAS	premises automation system
PLC	power line communication
SMPS	switch-mode power supply
THD	total harmonic distortion
WoT	Web-of-Things
WSN	wireless sensor network

REFERENCES

1. Remagnino, P. and Foresti, G.L. (2005) Ambient intelligence: a new multidisciplinary paradigm. *IEEE Transactions on Systems, Man, and Cybernetics – Part A: Systems and Humans*, 35 (1), 1–6.

2. Sivaneasan, B., Kumar, K.N., Tan, K.T., and So, P.L. (2015) Preemptive demand response management for buildings. *IEEE Transactions on Sustainable Energy*, 6 (2), 346–356.

3. Nguyen, N.-H., Tran, Q.-T., Leger, J.-M., and Vuong, T.-P. (2010) A real-time control using wireless sensor network for intelligent energy management system in buildings. *IEEE Workshop on Environmental Energy and Structural Monitoring Systems*, pp. 87–92.

4. Javed, A., Larijani, H., Ahmadinia, A., Emmanuel, R., Mannion, M., and Gibson, D. (2017) Design and implementation of a cloud enabled random neural network-based decentralized smart controller with intelligent sensor nodes for HVAC. *IEEE Internet of Things Journal*, 4 (2), 393–403.

5. Sivaneasan, B., Kumar, K.N., Tan, K.T., and So, P.L. (2015) Preemptive demand response management for buildings. *IEEE Transactions on Sustainable Energy*, 6 (2), 346–356.

6. Lee, S., Kwon, B., and Lee, S. (2014) Joint energy management system of electric supply and demand in houses and buildings. *IEEE Transactions on Power Systems*, 29 (6), 2804–2812.

7. Manic, M., Wijayasekara, D., Amarasinghe, K., and Rodriguez-Andina, J.J. (2016) Building energy management systems: the age of intelligent and adaptive buildings. *IEEE Industrial Electronics Magazine*, 10 (1), 25–39.

8. Guinard, D., Ion, I., and Mayer, S. (2011) *Search of an Internet of Things Service Architecture:RESTorWS-*? A Developers' Perspective (Proceedings of MobiQuitous)*.

9. Belimpasakis, P. and Moloney, S. (2009) A platform for proving family oriented RESTful services hosted at home. *IEEE Transactions on Consumer Electronics*, 55 (2), 690–698.

10. Wang, K., Wang, Y., Sun, Y., Guo, S., and Wu, J. (2016) Green industrial internet of things architecture: an energy-efficient perspective. *IEEE Communications Magazine*, 54 (12), 48–54.

11. Schachinger, D., Stampfel, C., and Kastner, W. (2015) Interoperable integration of building automation systems using RESTful BACnet Web services. *IECON 2015 – 41st Annual Conference of the IEEE Industrial Electronics Society*, pp. 003899–003904.

12. Corucci, F., Anastasi, G., and Marcelloni., F. (2011) A WSN-based Testbed for Energy Efficiency in Buildings. *Proceedings of the 16th IEEE Symposium on Computers and Communications (ISCC'11)*, pp. 990–993.

13. Wen, Y.-J. and Agogino, A.M. (2008) Wireless networked lighting systems for optimizing energy savings and user satisfaction. *IEEE Wireless Hive Networks Conference*, pp. 1–7.

14. Schoofs, A., Ruzzelli, A., and O'Hare, G. (2010) Appliance activity monitoring using wireless sensors. *Proceedings of the 9th International Conference on Information Processing in Sensor Networks, IPSN 2010, April 12–16, 2010, Stockholm, Sweden*.

15. Kim, W.H., Lee, S., and Hwang, J. (2011) Real-time energy monitoring and controlling system based on ZigBee sensor networks. *Procedia Computer Science*, 5, 794–797.

16. Yeh, L.W., Lu, C.Y., Kou, C.W., Tseng, Y.C., and Yi, C.W. (2010) Autonomous light control by wireless sensor and actuator networks. *IEEE Sensors Journal*, 10 (6), 1029–1041.

17. Guvensan, M.A., Taysi, Z.C., and Melodia, T. (2013) Energy monitoring in residential spaces with audio sensor nodes: TinyEARS. *Ad Hoc Networks*, 11, 1539–1555.

18. Sayed, K. and Gabbar, H.A. (2016) Scada and smart energy grid control automation, in *Smart Energy Grid Engineering*, Academic Press, pp. 481–514.

19. BeniniOpens, L., Farella, E., and Guiducci, C. (2006) Wireless sensor networks: enabling technology for ambient intelligence. *Mechatronics Journal*, 37 (12), 1639–1649.

20. Zhao, P., Suryanarayanan, S., and Simões, M.G. (2013) An energy management system for building structures using a multi-agent decision-making control methodology. *IEEE Transactions on Industry Applications*, 49 (1), 1–9.

21. Sun, B., Luh, P.B., Jia, Q.-S., Jiang, Z., Wang, F., and Song, C. (2013) Building energy management: integrated control of active and passive heating, cooling, lighting, shading, and ventilation systems. *IEEE Transactions on Automation Science and Engineering*, 10 (3), 588–602.

22. Barr, J. and Majumder, R. (2015) Integration of distributed generation in the Volt/VAR management system for active distribution networks. *IEEE Transactions on Smart Grid*, 6 (2), 576–586.

23. Missaoui, R., Joumaa, H., Ploix, S., and Bacha, S. (2014) Managing energy smart homes according to energy prices: analysis of a building energy management system. *Energy and Buildings*, 71, 155–167.

24. Lausten, J. (2008) *Energy Efficiency Requirements in Building Codes, Energy Efficiency Policies for New Buildings*, International Energy Agency, Paris, France.

25. China Energy Conservation Investment Corporation (2009) *China Energy Conservation and Emission Reduction Development Report (in Chinese)*, China Water Power Press, Beijing.

26. Tzempelikos, A. and Athienitis, A.K. (2007) The impact of shading design and control on building cooling and lighting demand. *Solar Energy*, 81, 369–382.

27. Moeseke, G., Bruyere, I., and Herde, A.D. (2007) Impact of control rules on the efficiency of shading devices and free cooling for office buildings. *Building and Environment*, 42, 784–793.

28. Xu, J., Luh, P.B., Blankson, W.E., Jerdonek, R., and Shaikh, K. (2005) An optimization-based approach for facility energy management with uncertainties. *HVAC&R Research*, 11 (2), 215–237.

29. Clarke, J.A., Cockroft, J., Conner, S., Hand, J.W., Kelly, N.J., Moore, R., O'Brien, T., and Strachan, P. (2002) Simulation-assisted control in building energy management systems. *Energy and Buildings*, 34, 933–940.

30. Ricquebourg, V., Menga, D., Durand, D., Marhic, B., Delahoche, L., and Loge, C. (2006) The smarthome concept: our immediate future. *1st IEEE International Conference on E-Learning in Industrial Electronics, December, Hammamet, Tunis*, pp. 1–5.

31. Wacks, K. (1993) The impact of home automation on power electronics. *Proceedings of Applied Power Electronics Conference and Exposition, March*, pp. 3–9.

32. Pan, M.S., Yeh, L.W., Chen, Y.A., Lin, Y.H., and Tseng, Y.C. (2008) A WSN-based intelligent light control system considering user activities and profiles. *IEEE Sensors Journal*, 8 (10), 1710–1721.

33. Jia, Q.-S., Shen, J.-X., Xu, Z.-B., and Guan, X.-H. (2012) Simulation-based policy improvement for energy management in commercial office buildings. *IEEE Transactions on Smart Grid*, 3 (4), 2211–2223.

34. Paracha, Z. and Doulai, P. (1998) Load management: techniques and methods in electric power system. *Proceedings of IEEE Energy Management and Power Delivery, March*, vol. 1, pp. 213–217.

35. Wacks, K. (1991) Utility load management using home automation. *IEEE Transactions on Consumer Electronics*, 37 (2), 168–174.

36. Mohsenian-Rad, A.H., Wong, V., Jatskevich, J., and Schober, R. (2010) Optimal and autonomous incentive-based energy consumption scheduling algorithm for smart grid. *Innovative Smart Grid Technologies (ISGT), January, Gothenburg, Sweden*, pp. 1–6.

37. Mohsenian-Rad, A. and Leon-Garcia, A. (2010) Optimal residential load control with price prediction in real-time electricity pricing environments. *IEEE Transactions on Smart Grid*, 1 (2), 120–133. (Pengwei, D. and Ning, L. (2011) Appliance commitment for household load scheduling. *IEEE transactions on Smart Grid*, 2, 411–419).

38. Ha, D. Long., Joumaa, H., Ploix, S., and Jacomino, M. (2012) An optimal approach for electrical management problem in dwellings. *Energy and Buildings*, 45, 1–14.

39. Paris, B., Eynard, J., Grieu, S., Talbert, T., and Polit, M. (2010) Heating control schemes for energy management in buildings. *Energy and Buildings*, 42, 1908–1917.

40. Clarke, J., Cockroft, J., Conner, S., Hand, J., Kelly, N., Moore, R., Brien, T., and Strachan, P. (2002) Simulation-assisted control in building energy management systems. *Energy and Buildings*, 34, 933–940.

41. Ha, D.L., Ploix, S., Zamai, E., and Jacomino, M. (2006) Tabu search for the optimization of household energy consumption. *IEEE International Conference on Information Reuse and Integration, Waikoloa, Hawaii, USA, September*, pp. 86–92.

42. Oldewurtel, F., Parisio, A., Jones, C., Gyalistras, D., Gwerder, M., Stauch, V., Lehmann, B., and Morari, M. (2012) Use of model predictive control and weather forecasts for energy efficient building climate control. *Energy and Buildings*, 45, 15–27.

43. Balan, R., Cooper, J., Chao, K., Stan, S., and Donca, R. (2011) Parameter identification and model-based predictive control of temperature inside a house. *Energy and Buildings*, 43, 748–758.

44. Ha, D.L., Ploix, S., Jacomino, M., and Le, M.H. (2012) Chapter 5, in *Home Energy Management Problem: Towards an Optimal and Robust Solution Energy Management*, INTECH, pp. 77–105.

45. Kelly, G.E., May, W.B., Kao, J.Y., and Park, C. (1994) Using emulators to evaluate the performance of building energy management systems. *ASHRAE Transactions: Symposium*, 100, 1482–1493.

46. Fanger, P.O., Stberg, A., Nicholl, A.G., Breum, N.O., and Jerking, E. (1974) Thermal comfort conditions during day and night. *European Journal of Applied Physiology and Occupational Physiology*, 33, 255–263.

47. Thevenard, D. and Haddad, K. (2006) Ground reflectivity in the context of building energy simulation. *Energy and Buildings*, 38, 972–980.

48. Lucas, F., Mara, T.A., Garde, F., and Boyer, H. (1998) A comparison between CODYRUN and TRNSYS, simulation models for thermal buildings behaviour. *World Renewable Energy Congress, Florence, Italy*.

49. Lee, E.-K. (2016) Advancing Building Energy Management System to Enable Smart Grid Interoperation, *International Journal of Distributed Sensor Networks*, 1–12.

50. Milam, M. and Venayagamoorthy, G.K. (2014) Smart meter deployment: US initiatives. *Innovative Smart Grid Technologies Conference (ISGT), IEEE PES ISGT 2014*, pp. 1- 5.

CHAPTER 3

SIMULATION-BASED ENERGY PERFORMANCE OF LOW-RISE BUILDINGS

FARAYI MUSHARAVATI,[1] SHALIGRAM POKHAREL,[1] and HOSSAM A. GABBAR[2]

[1]Department of Mechanical and Industrial Engineering, College of Engineering, Qatar University, Doha, Qatar
[2]Faculty of Energy Systems and Nuclear Science & Engineering and Applied Science, University of Ontario Institute of Technology, Oshawa, Canada

3.1 INTRODUCTION

Buildings are essential structures in our lives. They serve several needs of society such as providing shelter – a basic human need. Buildings also provide security, living space, and privacy to human beings. They also provide space to comfortably live and work in. As such, the performance of buildings is crucial to the development of any society.

The performance of a building or built environment is the efficiency at which the building functions and its impact on natural environment, urban environment, and its users. Engineers and scientists spent a lot of time putting value into the construction and operations of a building in a bid to improve efficiency. Buildings have a significant impact on the society. For example, it has been reported that buildings generally consume 40% of available energy [1]. In addition, buildings are responsible for 36% of CO_2 emissions [1]. This impact is even more pronounced in rapidly developing countries where the number and size of buildings in urban areas continue to increase resulting in an increased demand for energy in buildings. Moreover, buildings have a significant contribution to climate changes. As such, it is necessary to find methods, tools, and techniques for improving the performance of buildings.

Energy Conservation in Residential, Commercial, and Industrial Facilities, First Edition.
Edited by Hossam A. Gabbar.

The performance of buildings can be improved in two main ways, namely, (i) improving the energy use of buildings, and (ii) reducing the environmental impact of buildings. Reducing energy use in buildings can be achieved in many ways, namely, (a) by implementing energy saving measures and policies, (b) by implementing energy-efficient methods and techniques, and (c) by implementing energy conservation practices. However, all such improvements should be achieved without compromising thermal comfort of buildings. Reducing environmental impact of buildings can be done by reducing CO_2 emissions associated with the building and/or implementing onsite clean energy generation. It is, therefore, necessary to investigate the performance of buildings since information gained can be used to reduce the negative impact of buildings.

This chapter is about the energy performance of low-rise buildings. A low-rise building is an enclosed structure whose architectural height is below 35 m, and the structure is divided at regular intervals into occupiable levels [2]. However, a building is considered to be a construction as a whole, including its envelope and all technical building systems, for which energy is used to condition the indoor climate. Technical building systems include the technical equipment for heating, cooling, ventilation, domestic hot water, lighting, electricity production, and other building processes that require energy input.

Energy performance of a building refers to the amount of energy required to meet various needs associated with the use of a building [3–5]. For residential buildings, this includes energy required for heating, cooling, ventilation, lighting, and for using household appliances. In addition to this list, energy requirements for commercial buildings also include energy needed to operate building processes.

Energy performance of buildings usually takes many issues into account. In many cases, the energy performance of buildings is related to building insulation, technical and installation characteristics, design and positioning of buildings with respect to climatic conditions, solar exposure, own-energy generation (if any), type and nature of processes used in the building, and indoor climate control. It is therefore necessary for engineers to plan and design buildings in such a way that the energy performance of the resulting building will be acceptable and will have less negative impact on the society.

A number of factors contribute to energy use in buildings. For residential buildings, such factors include sociodemographics, attitudes, and self-reported behaviors as well as other building factors [6]. While electricity use in domestic buildings depends on occupants' need for energy services, such as light, comfort, and entertainment, overall energy use depends on a complex interaction of the factors mentioned above. The energy use pattern varies depending on the time of the day, month of the year, and in most cases from season to season. For commercial building factors that contribute to energy use include building uses and ownership, type and nature of technical building system, size and complexity of energy systems, energy system operations, as well as the type and nature of occupants' engagement. The energy use pattern also varies depending on the time of the day, month of the year, and in most cases from season to season.

One way of understanding the energy performance of buildings is to measure the energy efficiency of a building. Energy efficiency of a building refers to quantity of the specific energy consumption of that building in comparison to reference energy consumption benchmarks for that particular type of building under defined climatic conditions [7]. Measuring the energy efficiency of buildings helps in identifying opportunities and initiatives for reducing energy use in buildings without sacrificing comfort levels. It also helps in determining appropriate mechanisms for financing energy efficiency measures in buildings, as well as help in identifying appropriate legislative and policy tools that can be used to promote energy efficiencies.

For both new and existing buildings, simulation has been found to be a very useful tool for analyzing and understanding the energy performance of buildings. Over the years, modeling and simulation have become conventional methods, tools, and techniques that are used by engineers, building owners (managers), building tenants, and environment authorities to understand the behavior and characteristics of energy in buildings [8–10]. Engineers use building energy modeling and simulation to improve building designs and optimize energy use. Building owners are keen on providing the necessary energy requirements at minimum cost; hence, they can use building energy modeling and simulation data to determine their returns on investment. Moreover, information obtained from energy modeling and simulation can be used by building tenants to find ways of lowering their monthly bills. On the other hand, environment authorities can use building energy modeling and simulation data to determine the alternative use of energy that is associated with the least emissions. Therefore, the importance of building energy modeling and simulation can never be overemphasized.

3.2 SIMULATION OF BUILDING ENERGY PERFORMANCE

Due to the inherent need of building energy modeling and simulation, a number of technical models, emulators, prototypes, and stimulators have been developed for the purpose of analyzing energy use in buildings. Among others, prominent software tools for analyzing energy in buildings include BLAST, BSim, DeST, DOE-2.1E, ECOTECT, Ener-Win, Energy Express, Energy-10, EnergyPlus, eQUEST, ESP-r, IDA ICE, IES, HAP, HEED, PowerDomus, SUNREL, Tas, TRACE, and TRNSYS. In these tools, a large number of modeling features are available for designers and developers of building energy modeling and simulation applications. Such features include zone loads; building envelopes, daylighting, and solar requirements; infiltration, ventilation, and multizone airflows; renewable energy systems; electrical systems and equipment; HVAC systems; HVAC equipment; environmental emissions; economic evaluation; and climate data. The above-mentioned tools can be used in a variety of buildings, including residential, commercial, and industrial buildings and facilities. In the following sections, a description of the basic concepts and techniques in building energy modeling and simulation is provided, followed by sections on case studies that illustrate the significance of building energy modeling and simulation tools in assessing and evaluating energy use in buildings.

Building simulation involves the use of a computer to create a drawing or an image (2D or 3D) of a building for the purpose of studying the building. Most computer software tools (some of which have already been mentioned in the previous section) offer designers an environment with component parts of a building from which one can assemble the components into an image that closely represents the building under study. Usually data on the various components are available in the computer environment and the simulation can be carried out by linking the created image of the building with, for example, weather conditions for which the building simulation is required. The importance of building simulation lies in that it provides engineers and designers with a means of quantitatively predicting the future performance of a given building. For analytical purposes, building simulation focuses on building load design and energy analysis.

In buildings, a load is a source of energy demand and/or demand on cooling/heating equipment. Building energy design load is the maximum amount of energy that a building is designed to handle, that is, energy needs of a building. Such energy demand can be provided in various ways, including electricity, fuel, alternative energy, or by passive means. Loads are critical to building design because they determine the nature, type, and magnitude of activities that can be done within the building. Building energy design loads depend on (a) activities that will be done in the building, for example, nature, characteristics, and behavior of energy use, (b) size and shape of the building, and (c) spatial location of the building, which is influenced by climate (sun, clouds, wind, temperature, humidity, and precipitation) and site conditions (building's immediate surroundings). Loads consume energy and they influence the heat balance of a building. As such, understanding heat transfer mechanisms and how energy flows in a building is a crucial step in building energy load design.

In building simulations, load design is used to determine a number of issues including the various loads associated with a given building, volumetric airflow requirements, equipment capacities, and similarities and differences between equipment options for heating/cooling a space. On the other hand, building energy analysis is used to assess and evaluate the expected energy use in a given building. Such assessments provide a platform for the energy performance comparison of buildings during the early alternative design strategies. This includes predictions of the expected energy use of a proposed design and comparison of the effectiveness of building form, orientation, and envelope design options. Typical inputs include climate, envelope, internal gains from lighting, equipment, and occupants, and HVAC systems. In the following section, a case study is provided to illustrate the concept of building energy simulation.

3.3   CASE STUDY I: BUILDING ENERGY SIMULATION IN RESIDENTIAL BUILDINGS

The aim of this case study is to simulate the total cooling load profile for a residential building and evaluate the cooling performance of different cooling systems. The objectives are to (a) estimate the contribution of the internal loads (occupancy, light, equipment), (b) estimate the contribution of the external loads (ventilation, building envelope), and (c) suggest ways of improving the energy performance of typical residential units. In this case study, building energy simulation was conducted in a software known as HEED [11,12].

3.3.1   HEED

HEED is a Home Energy Efficiency Design tool that helps address issues related to building energy performance without compromising thermal comfort of building users. HEED addresses these issues at the very beginning of the design process. For existing buildings, HEED can be used to model the performance of the existing building and simultaneously suggest a more energy-efficient design. These suggestions can be used to retrofit an existing building, thereby improving the building's energy performance. HEED is also capable of analyzing the potential impact of building design decisions on human thermal comfort, energy costs, and environmental impact. As such, HEED can help stakeholders understand complex ideas about a building's performance. The software program, HEED, has been validated and its output can be used to make decisions without loss of generality [13]. In this case study, HEED is used to model the energy performance of an existing building in Qatar for the purpose of evaluating the building's energy performance.

3.3.2   Case Study Description

In this case study, the energy requirements of a typical residential building were simulated under the climatic conditions of Qatar. Figure 3.1 shows the outline design of the building.

For the case study house, all exterior walls and roofing are thermally insulated and all windows are double-glazed. The dominant material used for the construction of the house is red brick, which acts as one of the best insulators in the gulf region. The total number of windows in the house is 41. Windows are of two different sizes. The first type is relatively large with dimensions of (2000 mm × 1200 mm) and constitute a total of 30 windows. The other 11 windows are relatively smaller with dimensions of (1000 mm × 300 mm) and they are used in the toilets. There are two kinds of doors: two big front doors with dimensions of (2200 mm × 2757 mm) made of steel, and 20 wooden doors with dimensions of (2000 mm × 1000 mm).

3.3.2.1   *Building Load Profiles*   In buildings, the cooling load is the rate at which energy must be removed from a space to maintain the temperature and

FIGURE 3.1   Typical residential house in Qatar.

humidity at design values. Cooling is an essential component of any building that helps reduce the effects of heat gains in any occupied space in order to maintain the space comfort. Factors considered to be influencing internal cooling loads are occupancy, lighting, household equipment, and indoor temperatures. In the simulation of the cooling load profiles, two schemes were implemented in HEED software, that is, simulation of one house using district cooling system and another using conventional cooling system.

Figures 3.2–3.4 show plots of three internal loads (lighting, occupancy, and equipment) for the case study house during time of the day and over the whole year. The lighting load profile (Figure 3.2) shows two peaks: one around 6:00 a.m. (sharp peak) and the other around 6:00 p.m. (broader peak). It can also be observed, from Figure 3.2, that the lowest average lighting consumption is 1.09 kBtu/h, while the highest is 5.47 kBtu/h. For the monthly lighting consumption, it is clear from the graph that the lighting consumption is somewhat lower (4.38 kBtu/h) from March to August. If the building was to be designed to take advantage of natural daylight, this plot is expected to look like a U-shaped valley with the highest depression occurring when daylighting is greatest.

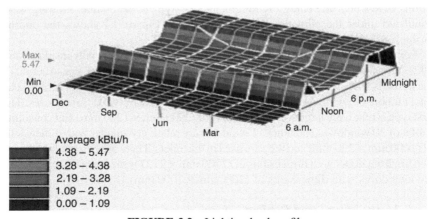

FIGURE 3.2   Lighting load profile.

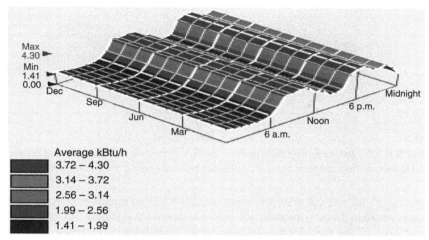

FIGURE 3.3   Occupancy load profile.

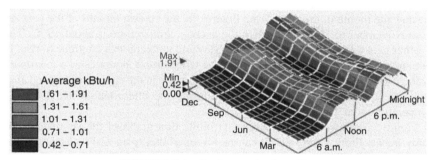

FIGURE 3.4   Equipment load profile.

The occupancy load profile is shown in Figure 3.3. This figure shows two major peaks: one occurring around 12:00 noon and the other around 12:00 midnight. The maximum consumption is about 4.30 kBtu/h, while the minimum is 1.41 kBtu/h.

The equipment load profile is shown in Figure 3.4. This figure represents heat generated inside the house envelope by all electrical and gas appliances used in the case study house. This figure shows a gradual increase in the use of equipment starting at 6:00 a.m. with a minimum of 0.42 kBtu/h and then increases as the occupants starts to wake up and use the house facilities. This energy consumption reaches a maximum of 1.91 kBtu/h by noontime.

Figure 3.5 shows two indoor temperature profiles that represent the cooling effects of two types of cooling load profiles, namely, district cooling (Figure 3.5a) and conventional cooling (Figure 3.5b) methods. Indoor temperatures when using district cooling ranges from 73.94 to 75 °F (i.e., 23.2–23.89 °C), while those for the conventional cooling system it ranges from 74.16 to 75.01 °F (i.e., 23.42–23.89 °C). However, it can be noted that these values are slightly higher than the ASHRAE

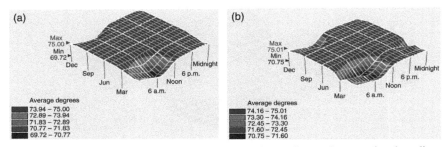

FIGURE 3.5 Indoor temperature profiles for district cooling and conventional cooling. (a) District cooling temperatures (°F). (b) Conventional cooling temperatures (°F).

standard 55 for thermal comfort, which is 22.5 °C. Both graphs exhibit the same shape, where in the early morning the indoor temperature is slightly lower than any other time of the day. Factors considered to be influencing external loads are ventilation, building envelope, and outdoor temperatures.

Figure 3.6 shows the variation of outdoor temperature with respect to time of the day and month of the year. From Figure 3.6, the coolest months of the year are from November to March for which the average temperature is about 19 °C. The weather gets warmer between April and May and temperatures continue to rise. In summers, that is, from June to October, the temperature increases to a maximum average of 42 °C. On an average day, the temperature in the early morning and after midnight is lower than the temperature during the afternoon, where the peak temperature is at 12 noon.

Figure 3.7 shows the ventilation and infiltration graph of the simulated case study house. From Figure 3.7, it can be observed that there is direct relationship between ventilation/infiltration and the cooling load. This is because of the air tightness of a building. Homes cannot be completely sealed from the outside environment. Instead, air enters or leaves the building from small cracks around

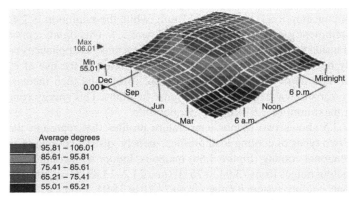

FIGURE 3.6 Outdoor temperature variations.

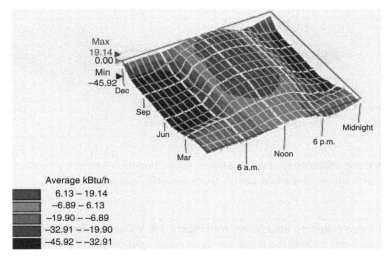

FIGURE 3.7   Cooling load profile and ventilation.

windows and doors, through vents around electrical outlets, and when people enter and/or leave the house. The maximum value of ventilation is 19.14 kBtu/h and it is from 9:00 a.m. to 4:00 p.m. During that period of time, the sun is perpendicular to the building, where maximum outdoor temperatures are often reached. In addition, the occupants, equipment, and lights are used most in that time. In general, a typically older home has 0.5 or more air changes per hour and newer tighter homes might have only 0.3 or less air changes per hour.

3.3.2.2 *Building Envelope Load Profiles*   Simulation of the building envelope was conducted for south-side windows, doors, and walls; north-side windows, doors, and walls; east-side windows, doors, and walls; and west-side windows, doors, and walls. Figure 3.8 shows the output of simulating the building envelope from the southern direction. The south-side building envelope consists of 12 windows, but there are no doors. The south-side building envelope provides the highest source of passive gain for the case study house since the angle of incidence between the beam radiation and the vertical south glass is closest to perpendicular at midday in winters. From Figure 3.8, it can be observed that the effect of the southern wall of the house on the cooling load profile is very low (0.85 Btu/h) in comparison to the effect of the southern windows (23.95 Btu/h). The southern window is the best source of passive gain for the case study.

As evident in Figure 3.8, the plot is saddle-shaped because it gains most of its heat in midday in winters when it is badly needed, and gains very little heat during summers when it is a liability. This happens because the angle of incidence between the beam radiation and the vertical south glass is closest to perpendicular at midday in winters. If these windows are shaded by an overhang, the spring and

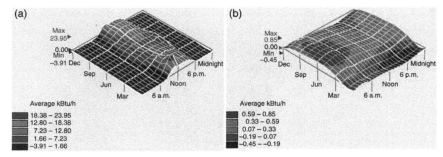

FIGURE 3.8 Simulation of the building envelope from the south side of the case study house. (a) South windows load profile. (b) South walls load profile.

fall will also gain very little heat. From Figure 3.8, it can be inferred that the effect of the southern wall of the house on the cooling load profile of the house is very low (0.85 Btu/h) compared to the effect of the southern window that is 23.95 Btu/h.

Figure 3.9 shows the output of simulating the building envelope from the northern direction.

The north side has five windows and one door. The effect of the northern wall of the house on the cooling load profile is about 1.59 Btu/h. Northern walls have higher effect on the cooling load than the southern walls and the heat gain to the building from that side is higher than the southern wall. The sun shines on northern walls in summers early in the morning just after sunrise, and then again late in the afternoon just before sunset. This is because between March 21 to September 21 (vernal and autumnal equinoxes), the sun rises to the north of east and sets to the north of west.

Figure 3.10 shows the output of simulating the building envelope from the eastern direction of the case study house.

As shown in Figure 3.10, the simulation results in a heat-mountain-shaped plot. Figure 3.11 shows the load profile for the west side of the case study.

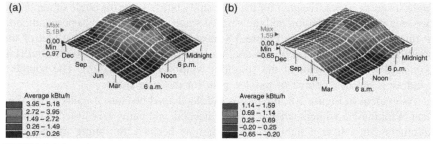

FIGURE 3.9 Simulation of the building envelope from the north side of the case study house. (a) North windows and doors load profile. (b) North walls load profile.

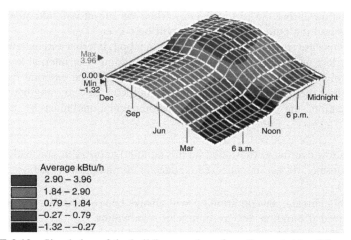

FIGURE 3.10   Simulation of the building envelope from the east side of the case study house.

Since the original orientation of the house is on the north–west, the house get its minimal radiation from the sun during the day, and the only time where the west windows get affected by the radiation is at the end of the day when time approaches sunset. The maximum is 3.60 kBtu/h during the day.

Simulation of external factors related to the building envelope shows that windows, doors, and walls that are on the eastern side of the house require higher energy demand for thermal comfort than any other side and this is because the solar radiation is very high during the day. In contrast, the western side windows, doors, and walls have the least demand across all the other sides and this is because the

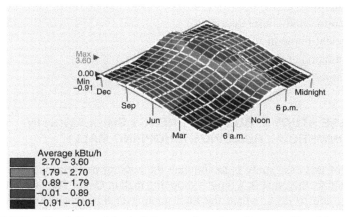

FIGURE 3.11   Simulation of the building envelope from the west side of the case study house.

solar radiation during the end of the day when the sunset will take place will be much less and the outdoor temperature will be cooler.

From this case study, it can be concluded that both internal and external factors have an effect on the energy performance of a building. For internal loads, lights have the highest effect on the cooling load profile. For the external factors, the orientation of the building can affect the cooling load. In designing houses in the given location, the following general recommendations must be taken into account:

- Window overhangs (designed for this latitude) or operable sunshades (extend in summer, retract in winter) can reduce or eliminate air-conditioning energy need.
- In this climate, air-conditioning will always be required, but can be greatly reduced if building design minimizes overheating.
- Raising the indoor comfort temperature limit will reduce air-conditioning energy consumption (raise thermostat cooling set point).
- Good natural ventilation can reduce or eliminate air-conditioning in warm weather, if windows are well shaded and oriented to prevailing breezes.
- Use plant materials (ivy, bushes, trees) especially on the west to shade the structure (if summer rains support native plant growth).

For new buildings in this location, designers should take into consideration many factors:

- Minimizing heat losses
- Maximizing solar heat gain
- Considering thermal mass
- Insulation
- Windows and conservatories
- Efficient home heating/cooling
- Ventilation
- Lighting and appliances

3.4 CASE STUDY II: BUILDING ENERGY SIMULATION IN COMMERCIAL BUILDINGS (SHOPPING MALL)

The aim of this case study is to simulate the cooling performance of a low-rise commercial facility, that is, a typical shopping mall in Qatar. The objectives of the simulation are to (a) compare the total cooling energy related to conventional cooling, unscheduled district cooling, and scheduled district cooling systems, (b) compare the energy performance related to each of the three cooling systems mentioned above, and (c) suggest ways of improving the energy performance of typical shopping malls.

For the sake of understanding cooling demands for a specific type of building, one should investigate the factors affecting the needed cooling energy. Many factors contribute to the amount of cooling load needed. Common factors in the public literature include building lighting, building ventilation, building occupancy, and most obviously the local area's weather conditions [14]. The building energy performance simulation was conducted in a software known as eQUEST [15].

3.4.1   eQUEST

The enhanced DOE-2.2-derived user interface, eQUEST program, was selected and used for the numerical simulation analysis. This program is most widely recognized and is able to perform detailed analysis and evaluate the required building technology and associated information for varied types of buildings [15]. Regional climate information with regard to the building's location and additional input data related to construction, operation, utility rate schedule, ventilation, and air-conditioning (HVAC) equipment are usually required to perform the simulation. Input to the software program eQUEST consists of a detailed description of the building under consideration. This includes information on the building's occupancy, building envelope, lighting, thermostat settings, and building equipment and processes. With eQUEST's parametric analysis, results from up to 10 cases can be compared on a single graph via eQUEST's results output mode [15].

Based on weather data for a specific location, eQUEST can simulate the long-term energy performance of a building. Hourly patterns of energy consumption over an entire year can be obtained, thus enabling decision makers to identify problem situations and hence find ways and means of minimizing energy consumptions. The program provides very accurate simulation of such building features and the dynamic response of differing air-conditioning system types and control. The program has been validated and its output can be used to make decisions without loss of generality [16].

3.4.2   Case Study Description

The case study mall occupies a total area of $128,000\,m^2$. It is a two-story high establishment housing with an array of stores and culinary venues. The mall consists of a total of 160 retail stores, 18 cafes and restaurants, and mall parking area totaling $1200\,m^2$ in the basement level and $800\,m^2$ in the ground-floor level. The mall holds a north–eastern orientation. All exterior walls and roofs are thermally insulated. The dominant building material is sand-blasted concrete. An estimate of 6–10% of the mall uses glass. Glass windows are double-paned, and argon, krypton, and xenon gases are used for insulation between the glass panes. The mall employs its own closed-loop district cooling system supported and maintained by a local company. As discussed in the introduction, a number of factors affect the energy performance of buildings. Some of the factors discussed in the following sections include occupancy engagements, lighting needs, ventilation factors, and building envelope factors.

3.4.3   Mall Occupancy

The number of visitors to the mall was observed in order to determine peak hours, the maximum number of visitors, as well as the patterns of the visits. Obtaining such information helps in determining the needed cooling load patterns according to the exhibited demand profile. The mall operates 7 days a week as follows: Saturdays to Thursdays from 10 a.m. to 10 p.m. and on Fridays between 2 p.m. and 10 p.m. However, some cafes in the mall might start operating earlier in the day where a few tenants from the surrounding residential areas are likely to visit the mall. Half-hourly observations of mall occupancy resulted in the patterns shown in Figure 3.12. As can be seen in Figure 3.12, the pattern of mall occupancy changes toward the weekend, that is, starting from Thursday to Saturday. It can be noted from Figure 3.12 that the number of recorded customers on Sunday was the lowest relative to other weekdays. During the week, the number of customers increases from Sunday to Saturday. As such, the cooling requirements and energy consumptions in the mall are expected to follow a similar trend.

3.4.4   Mall Lighting

There are mainly three types of lightings fixed throughout the mall. Mostly lightings are assumed to be equally dispersed as florescent and halogen. However, one restaurant implements LED lighting. The type of lighting used and the hours of lighting have a direct impact on the energy requirements of the mall.

3.4.5   Mall Ventilation

In the operations of the mall, there is a noticeable of movement of the air inside and outside the building due to several reasons. The mall has numerous entrances. The ground floor has five main entrances. Entrances 1 and 2 are located in front of the mall building. The other two entrances are placed at the back of the mall, that is, south and north. The fifth entrance is located on the eastern side. In addition to public entrances, there are a number of staff entries throughout the mall. One gate different from the others is one that leads outside to the outdoor stores and back into the mall. All public entrances employ automatic sliding doors that in turn permit movement of air out and into the building. In addition to the aforementioned entries, the mall accommodates lower floor entrances for the basement parking. These entries consist of six gates leading to the same assembly point that allows access to the mall's ground and first floors via escalators and elevators.

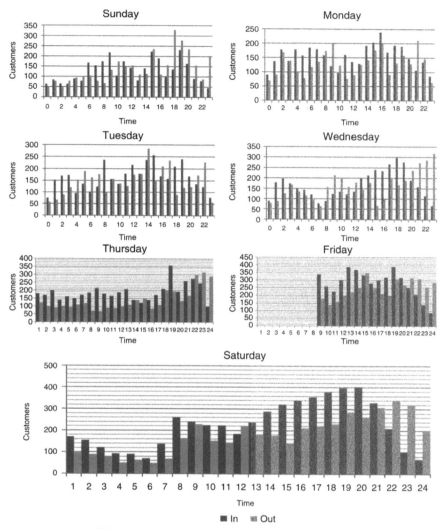

FIGURE 3.12 Mall occupancy for a typical week.

3.4.6 Mall Climate Control

The mall employs its own district cooling system, provided and maintained by a local company. District cooling is supplied to the mall through two main units: a primary unit that is supplied from the service provider, and a secondary unit that distributes cooling inside the mall. The main large pipes (supply and return) span a distance of 5 km from the provider. The mall's distribution network is a closed loop that contains an amount of 16,000 gallons of circulating water throughout the mall. Furthermore, heat exchangers are used as an interface between provider pipes and mall pipes. The heat exchangers exchange energy not water supply. The secondary station at the mall is mainly composed of four pumps, actuators, motors, chilled water pipes, a chemical dosing tank, a heat exchange, and a water filter (strainer). Under normal conditions, three pumps operate while the fourth is on standby in case one should fail. However, pumps (including the standby pump) are run on variable frequency speed drives that run the pumps higher or lower revolutions per minute if needed when pressure drops. In addition, every store in the mall has its individually installed thermostat. The thermostats are used to detect the space's temperature and adjust the need for cooling accordingly.

The energy performance of buildings is of great significance, especially to building owners, for it ultimately translates to costs. Energy-efficient buildings are henceforth crucially desirable. For the mall, an investigation of building energy performance and comparisons of cooling systems was conducted, via simulation software. Through the software eQUEST, building energy performances were simulated for the mall in three scenarios:

a. *Conventional Cooling System*
   The first scenario assumes the building to be employing a traditional cooling system with its typical components and characteristics.
b. *District Cooling System*
   A simulation of the current system in the mall (nonscheduled district cooling system) with all field measurements collected as input parameters.
c. *Scheduled District Cooling System*
   A recommended schedule for the current system based on the occupancy profiles described earlier.

At the time of the study, the cooling system operated for 24 h per day. From demand profiles, operating 24 h a day is not necessary and is costly in terms of energy consumption. A cooling schedule based on occupancy would make more sense in terms of reducing the energy required for cooling purposes. A representative occupancy profile based on averaging the mall occupancy is shown in Figure 3.13.

In order to complete the simulation, field measurements were collected and used in entering data into the eQUEST software's input screens. In order to complete the creation of the desired building shell, a total number of 47 screens requiring detailed building information have to be filled according to the actual building's characteristics. Input data include comprehensive information about total and floor building areas, building shape and orientation, ventilation, lighting (including specific information on skylights), activity regions, their areas and occupancies, operation hours, and construction materials. Once all data have been entered and the building shell information is completed without error, the Building Creation Wizard will create the building's simulated geometry and orientation. After

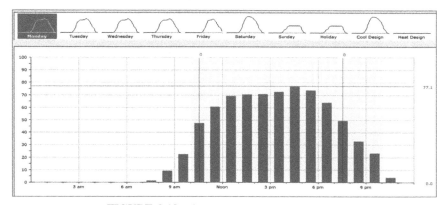

FIGURE 3.13 Occupancy profile in eQUEST.

developing the required building shell, the internal and external loads are specified and so should be the details and types of cooling systems.

Figure 3.14 shows a comparison of the monthly cooling energy consumption for the three systems: *conventional cooling system, unscheduled district cooling system,* and *scheduled district cooling system.* From Figure 3.14, the significant energy saving effect of using district cooling is quite apparent. In addition, more savings can be obtained if scheduled district cooling is implemented. More specifically, district cooling reduces the amount of energy needed for cooling to about 49%, while scheduled district cooling further reduced energy requirements by 14%.

Figure 3.15 shows comparison of the monthly energy consumption for three systems: *conventional cooling system, unscheduled district cooling system,* and *scheduled district cooling system.* From Figure 3.15, the significant reduction in energy consumption when using district cooling is quite apparent. It can be

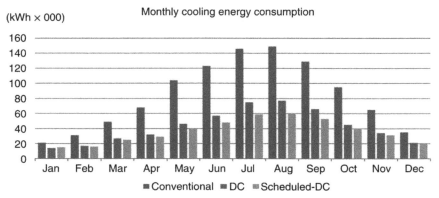

FIGURE 3.14 Comparison of monthly cooling consumption by end-use.

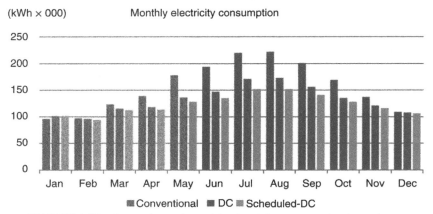

FIGURE 3.15 Comparison of monthly electricity consumption by end-use.

observed that a total of 16.4% in annual electricity consumption saving can be accomplished by using district cooling and a further 6.3% saving in electricity consumption is expected when the district cooling is scheduled to match mall occupancy.

More insight into the energy performance of the mall can be obtained by simulating the monthly peak cooling demand for each system. Understanding peak load demands is important for the design of a cooling system, especially for commercial buildings like shopping malls. When peak loads are present, rates for electricity usage during that time are usually higher. Furthermore, peak loads determine a greater percentage of the energy that the building is going to use. Peak loads obtained from the simulation run results for each of the three systems are shown in Figures 3.16–3.18. As shown in these figures, eQUEST software calculates the peak cooling load and integrates it with the cooling performance. Such information can be used to redesign buildings and building processes and equipment in a bid to save and conserve energy consumption.

It can be inferred from the simulation analysis in this section that a district cooling system is far more energy-efficient and cost-effective than the conventional cooling systems. Since a district cooling system is already applied in the mall, further observations revealed that the district cooling system employed lacks control and operational planning. Hence, there is room for further improvements.

The following are some suggestions generated to improve the buildings' energy efficiency:

a. *The Implementation of a Schedule*
   Although additional control implementation may be costly at the beginning, on the long run a scheduled system will save energy-consumption-related costs. Taking into consideration the mall's occupancy patterns will also help to ensure a comfortable atmosphere when needed.

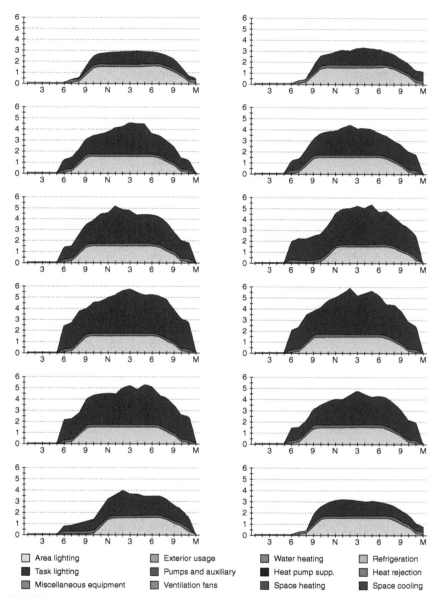

☐ Area lighting	▨ Exterior usage	▨ Water heating
■ Task lighting	■ Pumps and auxiliary	■ Heat pump supp.
▨ Miscellaneous equipment	☐ Ventilation fans	■ Space heating

▨ Refrigeration
▨ Heat rejection
■ Space cooling

FIGURE 3.16 Monthly electricity peak day profile for the conventional cooling system from January to December.

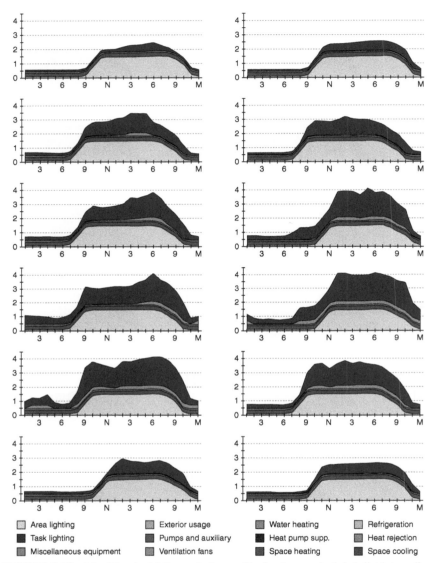

FIGURE 3.17 Monthly electricity peak day profile for the unscheduled district cooling system from January to December.

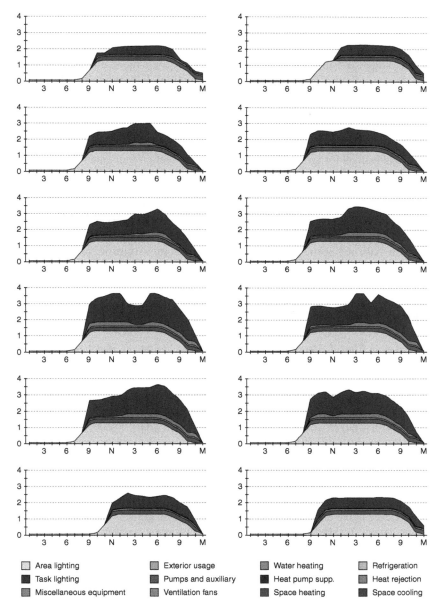

Area lighting	Exterior usage	Water heating	Refrigeration
Task lighting	Pumps and auxiliary	Heat pump supp.	Heat rejection
Miscellaneous equipment	Ventilation fans	Space heating	Space cooling

FIGURE 3.18 Monthly electricity peak day profile for the scheduled district cooling system from January to December.

b. *LED Lights*

It was noticed during our field measurements collection that most lights dispersed throughout the facility are of halogen and florescent bulbs type. It is suggested that these bulbs be replaced with more energy-saving and energy-efficient bulbs and lamps. For example, light-emitting diode (LED) lamps have been proven to be more efficient while far less in electricity usage than the halogen and florescent bulbs types used within the mall.

c. *Air Curtains*

Building ventilation is a factor that eventually affects the amount of cooling load needed for the building. In order to prevent air movement from and into the mall, it is suggested to make use of air curtains on every entrance. Entrances to the mall employ automatic sliding doors. These will allow a change in air every time they open. However, with the application of air curtains, this problem can be mitigated. Air curtains work as shields made of air pushed to the ground that will help prevent outside air from moving into the mall, and keep inside air from being released to outside. Additionally, air curtains would also help keep out undesired insects, dust particles, and unwanted odors.

REFERENCES

1. European Commission (2016) Energy efficiency in buildings. Available at https://ec. europa.eu/energy/en/topics/energy-efficiency/buildings (retrieved October 2016).
2. Emporis Standards (2015) Data Standards: Structures – Low-Rise Building. Available at https://www.emporis.com/building/standards/49213 (retrieved June, 2015).
3. Energy Performance of Buildings Directive 2002/91/EC. Available at http://eur-lex.europa.eu/LexUriServ/LexUriServ.do?uri=OJ:L:2003:001:0065:0071:EN:PDF (retrieved April 11, 2015).
4. The Energy Performance of Buildings (Certificates and Inspections) (England and Wales) Regulations 2007, March 23, 2007. Available at http://www.legislation.gov.uk/uksi/2007/991/pdfs/uksi_20070991_en.pdf (retrieved September 19, 2015).
5. Department for Communities and Local Government (2012) Improving the energy efficiency of our buildings: a guide to display energy certificates and advisory reports for public buildings. Available at https://www.gov.uk/government/publications/display-energy-certificates-and-advisory-reports-for-public-buildings (retrieved September 2015).
6. Huebner, G.M., Hamilton, I., Chalabi, Z., Shipworth, D., and Oreszczyn, T. (2015) Explaining domestic energy consumption: the comparative contribution of building factors, socio-demographics, behaviours and attitudes. *Applied Energy*, 159, 589–600.
7. Energy efficiency in buildings. http://www.unido.org/fileadmin/import/83276_Module19.pdf (retrieved September 2015).
8. Ayoub, N., Musharavati, F., Pokharel, S., and Gabbar, H.A. (2014) Energy consumption and conservation practices in Qatar: commercial building case study energy and buildings. *Energy and Buildings*, 84, 55–69.

9. Al-Awainati, N., Fahkroo, M.I., Musharavati, F., Pokharel, S., and Gabbar, H.A. (2013) Evaluation of thermal comfort and cooling performance of residential buildings in arid climates. *2013 IEEE International Conference on Smart Energy Grid Engineering (SEGE)*, pp. 1–6.

10. Fahkroo, M.I., Al-Awainati, N., Musharavati, F., Pokherel, S., and Gabbar, H.A. (2013) Operations optimization towards high performance cooling in commercial buildings. *2013 IEEE International Conference on Smart Energy Grid Engineering (SEGE)*, pp. 1–6.

11. Milne, M. and Kohut, T. (2002) Using HEED to design energy efficient affordable housing. *Proceedings of the American Solar Energy Society Conference, Reno, Nevada.*

12. Milne, M., Morton, J., and Kohut, T. (2006) Energy efficient affordable housing: validating HEED's predictions of indoor comfort. *Proceedings of the American Solar Energy' Society, Denver, CO.*

13. Murray, M. (2012) HEED Validation Reports.

14. Deng, S.-M. and Burnett, J. (2000) A study of energy performance of hotel buildings in Hong Kong. *Energy and Buildings*, 31.1, 7–12.

15. Ke, M.-T., Yeh, C.-H., and Jian, J.-T. (2013) Analysis of building energy consumption parameters and energy savings measurement and verification by applying eQUEST software. *Energy and Buildings*, 61, 100–107.

16. Rallapalli, H.S. (2010) A comparison of EnergyPlus and eQUEST whole building energy simulation results for a medium sized office building. Dissertation, Arizona State University.

PART II

ENERGY SYSTEMS

CHAPTER 4

FAST CHARGING SYSTEMS

HOSSAM A. GABBAR[1,3] and AHMED M. OTHMAN[1,2]
[1]Faculty of Energy Systems and Nuclear Science, University of Ontario
Institute of Technology, Oshawa, Canada
[2]Electrical Power and Machines Department, Faculty of Engineering, Zagazig
University, Zagazig, Egypt
[3]Faculty of Engineering and Applied Science, University of Ontario Institute of
Technology, Oshawa, Canada

4.1 INTRODUCTION

Flywheel kinetic energy storage offers very good features such as power and energy density. Moreover, with some different range vehicles, this technology can be enough to supply all the energy to the power train. The challenges to be met to integrate such technology in vehicles are the mass, the efficiency, and especially the cost. In this chapter, a technoeconomic optimization of a flywheel energy storage system is presented. It is made up of a flywheel, a permanent magnet synchronous machine, and a power converter. For each part of the system, physical and economical models are proposed. Finally, an economic optimization is done on a short-range ship profile.

Flywheel energy storage has become one of the important energy storage in the world. That is why flywheels energy storage are used a lot in power systems and microgrid recently as they are flexible, smart, and active. In addition, they fit more with renewable resources and are considered to be environment-friendly.

Quick charging of modern energy storage has turned into a standard charging innovation because of the operational reserve funds, expanded profitability, and well-being that this innovation offers. Clients have understood the advantages of quick charging and proceed to understand the advantages at assembling plants and dissemination focuses all through many applications.

This chapter proposes a control system for module with flywheel-based fast charging system (FFCS). The fundamental part of the FFCS is to trade off the

Energy Conservation in Residential, Commercial, and Industrial Facilities, First Edition.
Edited by Hossam A. Gabbar.

predefined charging profile of battery and incorporated the arrangement of a hysteresis dynamic power converter supported to the power framework.

In this sense, when the dynamic power is not being separated from the network, FFCS gives the power required to maintain the persistent charging procedure of battery. A key trademark of the entire control framework is that it can work without any discrete correspondence between the frameworks tied and FFCS converters.

4.2   FAST CHARGING VERSUS OTHER CHARGING APPROACHES

An electric vehicle charging station, also called electric reviving point, charging point is a component in a foundation that has provisions for electrical energy charge for the energizing of electric vehicles, for example, module electric vehicles. Most of the charging platforms are on-road locations associated with electrical service networks and others are situated within retail commercial malls or to be worked through numerous privately owned businesses (Table 4.1).

Charging stations are categorized into four main groups:

1. *Private Charging Stations:* An electrical vehicle (EV) connects to charger when returns home and then auto recharge through nighttimes. A private station ordinarily does not need any client validation nor any metering system; it just needs simple installation for a devoted circuit. Other convenient chargers will likewise need partition installation as in stations.

2. *Public Stations:* Business scheme for an expense or without fees that is serviced in organization through the holders of certain parking area. The charging process might be a moderate mode that urges vehicles holder to charge the autos during shopping and the like. It can involve shopping malls and centers or a business' own particular representatives.

3. *Fast or Quickly Charging:* For charging demand greater than 40 kW, conversing more than 120 km within time frame of 10 and 25 min. They will likewise need routine utilization by suburbanites at metropolitan places and to be charged when stopped for short and/or long time frames.

4. *Swaps or Changes:* Process for batteries in less than 15–20 min. Predetermined focus on zero- and low-emission vehicle and its range is less than 15 min. It can be done with vehicle battery by swaps and also in hydrogen

TABLE 4.1   Gas versus Electric Charging

Vehicle	Charge as Per 100 km	Cost per Unit	Cost
Electric	16 kWh	× $0.08/kWh	= $1.28
Gasoline	8.4 L	× $1.35/L	= $11.40

As another comparison, fuel = $0.13 per km, EV = $0.02 per km.
Electricity is cheaper nine times more than gasoline.
Electricity saves around $2000–3000 a year per car, based on 30,000 km/year.

TABLE 4.2 Charging Time of Different Charging Categories

Considering 100 km, Charging Time	Voltage (V ac)	Max. Current (A)	Power (kW)
6–8 h	220	16–18	3.2
3–5 h	220	32–36	7.4
2–3 h	400	16	10
1–2 h	400	32	22
20–30 min	400	63	43
20–30 min	450–505	120–135	50
10 min	350–550	320–380	120

fuel cell vehicles. This works in coordination with the request to refill from customary holders (Table 4.2 and Figures 4.1 and 4.2).

A critical increment of EVs in developing control markets would mean a development in power request giving some comfort to utilities working in energy markets, and likewise helping framework administrators adjust control-free market activity by means of vehicle-to-network ventures. The amassing of network-associated EV batteries would likewise energy boost development by smoothing supply irregularity and encouraging their incorporation into the power network. As shown, there is high power demand for EV, which confirms the need for fast charging infrastructures.

Fast charging versus other charging approaches

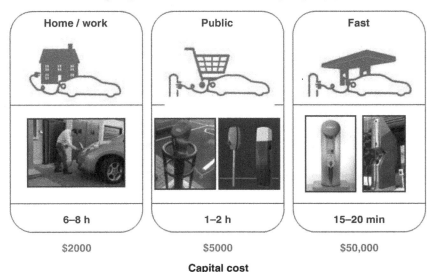

FIGURE 4.1 Fast charging versus other charging approaches.

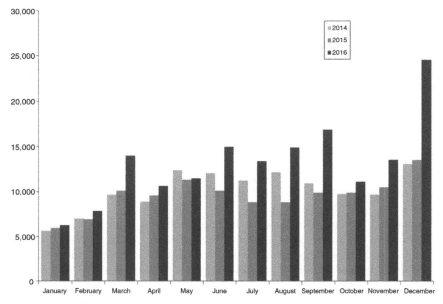

FIGURE 4.2   Sales trend of electric vehicles till end of 2016, USA.

4.3 FAST CHARGING: TECHNOLOGIES AND TRENDS

DC fast chargers replace level 1 and level 2 charging stations, and are intended to charge electric vehicles rapidly with an electric yield extending between 50 and120 kW. Most present-day wholly electric vehicles can be charged with DC fast charge capacity, and there are about 2200 rapid chargers now fit for adding huge range to an EV in a very little longer time than it take to fill your gas tank.

Fast charging can be achieved using level 2 (basic) or ultimate (level 3) charging stations:

AC level 1:   117 V 16 A max (normal charging)
AC level 2:   240 V 32 or 70 A (basic fast charging)

Charger specifications level 3 (ultimate: direct DC):

Input   Three-phase 200 V
Output: Max DC 45–50 kW (can reach 200 kW)
           Max DC voltage 700 V
           Max DC current 750 A

Fast charging is advantageous to demand charge distribution network in big cities that is very congested, and peak demand reduction is a matter of concern. These advantages include grid services to provide peak power, by turn peak load into base load. In addition, grid stabilization can be achieved by grid expansion and maintaining promised levels of supply.

4.3.1 Flywheel Technology

Flywheel energy storage systems utilize active kinetic rotation that can be stored in the rotated mass with low levels of losses in the friction component (Figure 4.3). The input energy such as electrical increases the acceleration of the mass to speed by means of an incorporated engine generator. This energy will be stored and depend on the moment of inertia and speed of the rotating shaft, and can be expressed as follows:

Kinetic energy: $E_k = \frac{1}{2}I\omega^2$, I: moment of inertia $I = \int r^2 dm$.

- For a cylinder, the moment of inertia $I = \frac{1}{2}r^4\pi a\rho$.
- Energy is increased if ω increases or if I increases.
  The optimization of the energy related to mass can be achieved by spinning flywheel with maximum possible speed.
- Max. speed: $v_{max} = \sqrt{\dfrac{2K\sigma_{max}}{\rho}}$

4.3.2 Advantages of Flywheel

1. *Higher Power Densities:* Especially for EV and HEV applications, flywheel is significant to supply much power density than battery (Figure 4.4).
2. *High Reliability and Cycle Life:* The power from flywheels is not degraded, and remarkable by reliability and high cycle life.
3. *Environmental Impact:* The environmental impact of flywheels is so friendly that there is no hazardous substance than can be recycled.

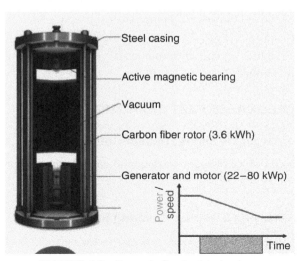

FIGURE 4.3 Example flywheel technology.

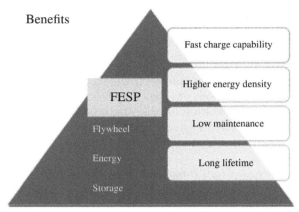

FIGURE 4.4 Benefits of flywheel technology (FESP: Flywheel Energy Storage Platform).

4. *Temperature Sensitivity:* Flywheels are for the most part less delicate to the surrounding temperature than batteries. Some essential warm breaking points are set by the temperature in the windings to abstain from softening; the magnets, to maintain a strategic distance from demagnetization; and the composite material, to abstain from blazing it.

4.3.3 Scalable Flywheel Technology

- Minimal frictional losses
- Advanced material: carbon-fiber-reinforced plastic
- Noncontact bearings
- Ultimate strength and speed up to 45,000 rev/min
- Significant charging cycles keeping raised up service life
- A wider temperature range
- Turnkey solution – kilowatt to multiwatt
- Remote controlled

4.4 FLYWHEEL-BASED FAST CHARGING SYSTEM

4.4.1 Fast Charging Stations: Design Criteria

The design criteria of a fast charging station to cover both residential and public charging infrastructures, FFCS.

4.4.2 Fast Charging Stations: Covering Factor

Fast charging stations can achieve a covering factor up to 95% of total electric charging demand. This can be achieved by three factors: (i) waiting time at the

FFCS; (ii) number of empty charging spots per FFCS per time; and (iii) maximum number of required charging spots per FFCS.

EVs can be charged wherever there is an electrical attachment. This includes open carports, auto parks, workplaces, grocery stores, healing centers, lodgings, homes of companions and colleagues, shopping centers, and neighborhoods. The increase of FFCS installations can replace some of the above electric charging options. This can be achieved by planning optimization and control (Figure 4.5).

4.4.3 Mobility Behavior

In MCD model, the mobility and charging demand of EVs is determined depending on the mobility behavior.

Designing a fast charging station starts mainly by studying the profile of the concerned charging load (these include social and technical analyses):

- *Social:* The population of the assigned region and how frequent they charge their EVs.
- *Technical:* The limits of the charging infrastructure.

The second step is the modeling that performs simulation of the real quick charging station within various scenarios.

4.4.4 Mobility Integrated Study

The starting point for the mobility model can be with 50 groups of 1500 vehicles each. Each vehicle is controlled in its charging based on the behavior of the owner:

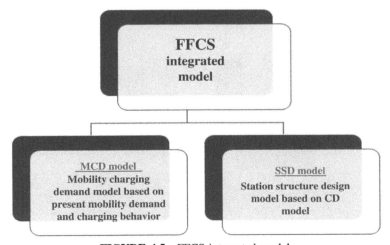

FIGURE 4.5 FFCS integrated model.

- The case of the EV as per its running or stopping (μv)
- The amount of consumed power from the electricity (μm)
- The places where the car usually stands at (μc), whether it is house places, working place, or others.
- Vehicle model (μt), the sizing of the EV.

4.5 FFCS Design (Figure 4.6)

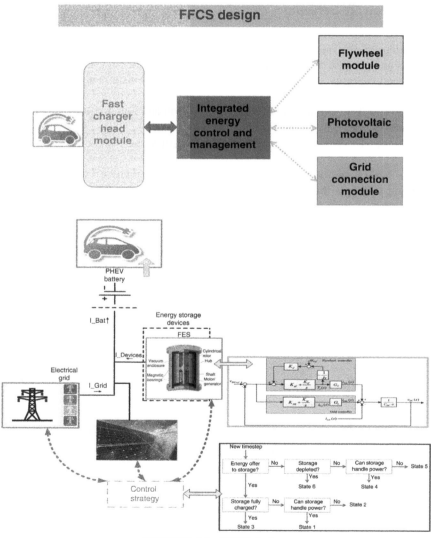

FIGURE 4.6 FFCS design.

4.5.1 FFCS: Multilevel Circuit Design

Multi/three-phase two-level AC/DC converter as grid interface and flywheel converters is shown in Figure 4.7; it depends on fast DC/DC converter to quick charging where each component is installed by connection via a common DC bus.

4.5.2 Control of Flywheel by Hysteresis Controller

The points of interest in utilizing hysteresis control are significant dynamic response and capacity to control the peak demand and current swell in assigned hysteresis band range. It is likewise extremely successful for physical execution.

Likewise, it diminishes the harmonics, where it takes a shot as essential by identifying consonant current to compute the measure of the compensated current required for sustaining back to the power framework.

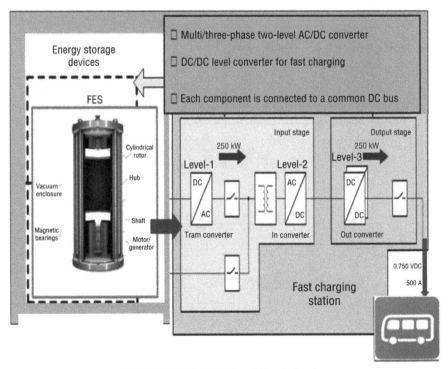

FIGURE 4.7 FFCS multilevel circuit.

4.6 PROPOSED SYSTEM DESIGN

The proposed FFCS (Figure 4.8) will be designed based on the following specifications:

- Develop and deploy FFCS in one (or more) selected electric bus route in cities.
- Install FFCS at suitable locations within the bus route, which will be selected with performance optimization algorithm.

FFCS will be remotely controlled and performance will be optimized with intelligent algorithms based on key performance indicators, including charging time, costs, efficiency, and demand coverage for other EVs. The design of FFCS is presented based on flywheel energy storage platform (FESP), which is presented to implement fast charging for transportation infrastructures (Figure 4.9).

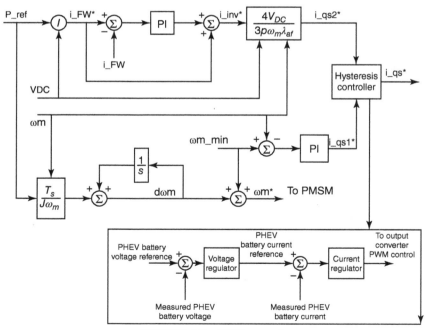

FIGURE 4.8 FFCS control.

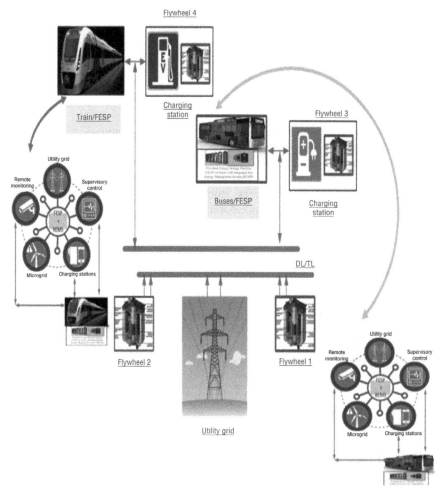

FIGURE 4.9   FFCS for energy-efficient transportation buses and railway.

4.7   ROI AND BENEFITS OF FFCS

The benefits and return on investment (ROI) ratio between normal charging and FFCS range between three and six times:

- *With respect to production within same time frame:*
    For X kWh of charging production, in time t.
- The amount of normal charging $= X$ kWh
- The amount of FFCS charging $= 3X$ kWh

- Considering saving of time reduction of FFCS:
  For fixing X kWh of charging production:
  - The production time of normal charging $= 3t–6t$
  - The production time of FFCS charging $= t$

4.8 CONCLUSIONS

The proposed FFCS is analyzed in view of user and technology requirements, where fast charging is proposed based on flywheel technology. Advantages of fast charging over traditional charging is expressed, in particular when dealing with congested cities where short charging time is critical. The design of FFCS is presented based on flywheel energy storage platform, which is presented to implement fast charging for transportation infrastructures: implementation schemes for eBuses and EVs.

FURTHER READINGS

Abapour, S., Abapour, M., Khalkhali, K., and Moghaddas-Tafreshi, S.M. (2015) Application of data envelopment analysis theorem in plug-in hybrid electric vehicle charging station planning. *IET Generation, Transmission, & Distribution*, 9 (7), 666–676.

Anglani, N. and Muliere, G. (2010) Analyzing the impact of renewable energy technologies by means of optimal energy planning. *9th International Conference on Environment and Electrical Engineering (EEEIC), Prague*, pp. 1–5.

Aubry, J., Ben Ahmed, H., and Multon, B. (2012) Sizing optimization methodology of a surface permanent magnet machine-converter system over a torque-speed operating profile: application to a wave energy converter. *IEEE Transactions on Industrial Electronics*, 59 (5), 2116–2125.

Azizipanah-Abarghooee, R. (2013) A new hybrid bacterial foraging and simplified swarm optimization algorithm for practical optimal dynamic load dispatch. *International Journal of Electrical Power & Energy Systems*, 49, 414–429.

CEMAC, Chung, D., Elgquist, E., and Santhanagopalan, S. *Automotive lithium-ion cell manufacturing: regional cost structures and supply chain considerations*, s.l. NREL, April 2016. NREL/TP-6A20-66086.

Cornic, D. (2010) Efficient recovery of braking energy through a reversible DC substation. *Electrical Systems for Aircraft, Railway and Ship Propulsion (ESARS)*, doi: 10.1109/ESARS.2010.5665264

Cossent, R. and Gomez, T. (2009) Towards a future with large penetration of distributed generation: Is the current regulation of electricity distribution ready? Regulatory recommendations under a European perspective. *Energy Policy*, 37 (3), 1145–1155.

Fadaee, M. and Radzi, M. (2012) Multi-objective optimization of a stand-alone hybrid renewable energy system by using evolutionary algorithms: a review. *Renewable and Sustainable Energy Reviews*, 16 (5), 3364–3369.

Falvo, M.C., Lamedica, R., Bartoni, R., and Maranzano, G. (2011) Energy management in metro-transit systems: an innovative proposal toward an integrated and sustainable urban mobility system including plug-in vehicles. *Electric Power Systems Research*, 81 (12), 2127–2138.

Fan, P., Sainbayar, B., and Ren, S. (2015) Operation analysis of fast charging stations with energy demand control of electric vehicles. *IEEE Transactions on Smart Grid*, 6 (4), 1819–1826.

Gulin, M., Vašak, M., and Baotić, M. (2015) Analysis of microgrid power flow optimization with consideration of residual storages state. *Proceedings of the 2015 European Control Conference, Austria*, pp. 3131–3136.

Hong, Y., Wei, Z., and Chengzhi, L. (2009) Optimal design and techno-economic analysis of a hybrid solar-wind power generation system. *Journal of Applied Energy*, 86 (2), 163–169.

Leung, P.C.M. and Lee, E.W.M. (2013) Estimation of electrical power consumption in subway station design by intelligent approach. *Applied Energy*, 101, 634–643.

Liu, X., Wang, P., and Loh, C. (2011) A hybrid AC/DC microgrid and its coordination control. *IEEE Transactions on Smart Grid*, 2 (2), 278–286.

Musio, M., Serpi, A., Musio, C., and Damiano, A. (2015) Optimal Management Strategy of Energy Storage Systems for RES. *IECON 2015: 41st Annual Conference of the IEEE Industrial Electronics Society*, pp. 5044–5049.

Okui, A., Hase, S., Shigeeda, H., Konishi, T., and Yoshi, T. (2010) Application of energy storage system for railway transportation in Japan. *International Power Electronics Conference (IPEC)*, June, pp. 3117–3123.

Pankovits, P., Pouget, J., Robyns, B., Delhaye, F., and Brisset, S. (2014) Towards railway-smartgrid: energy management optimization for hybrid railway power substations. *IEEE PES Innovative Smart Grid Technologies Conference Europe (ISGT-Europe)*, Oct. 2014, pp. 12–15.

Patnaik, S.S. and Panda, A.K. (2012) Particle swarm optimization and bacterial foraging optimization techniques for optimal current harmonic mitigation by employing active power filter. *Journal of Applied Computational Intelligence and Soft Computing*, 2012, 1–10.

PNNL, Sandia, Conover, D.R., Ferreira, S.R., Crawford, A.J., Schoenwald, D.A., Fuller, J., Rosewater, D.M., Gourisetti, S.N., and Viswanathan, V. (2016) Protocol for Uniformly Measuring and Expressing the Performance of Energy Storage Systems. s.l.: PNNL, Sandia, SAND2016-3078 R, PNNL-22010 Rev.

Rajabi, A., Fotuhi, M., and Othman, M. (2015) Optimal unified power flow controller application to enhance total transfer capability. *IET Generation, Transmission & Distribution*, 9 (4), 358–368.

Salas, V., Olias, E., Alonsob, M. and Chenlo, F. (2008) Overview of the legislation of DC injection in the network for low voltage small grid-connected PV systems in Spain and others. *Renewable & Sustainable Energy Reviews*, 12 (2), 575–583.

Xinggao, L., Yunqing, H., Jianghua, F., and Kean, L. (2014) A novel penalty approach for nonlinear dynamic optimization problems with inequality path constraints. *IEEE Transactions on Automatic Control*, 59 (10), 2863–2867.

Yiqiao, C. and Jiahai, W. (2013) Differential evolution with neighborhood and direction information for numerical optimization. *IEEE Transactions on Cybernetics*, 43 (6), 2202–2215.

CHAPTER 5

MICROINVERTER SYSTEMS FOR ENERGY CONSERVATION IN INFRASTRUCTURES

HOSSAM A. GABBAR,[1] JASON RUNGE,[1] and KHAIRY SAYED[2]
[1]Faculty of Engineering and Applied Science, University of Ontario Institute of Technology, Oshawa, Canada
[2]Faculty of Engineering, Sohag University, Sohag, Egypt

5.1 INTRODUCTION

This chapter proposes and evaluates different methods of increasing the resiliency for microinverter systems in PV residential applications. The design of three inverters of different power levels is discussed. A preliminary comparison is carried out between the designed inverters and those currently available on the market. These designed inverters are applied into different scenarios to evaluate microinverter system resiliency.

In the companion study, three inverter designs with different power ratings were presented. First, the subcircuits and their respective mathematical models were shown, along with the PSIM models created for them. These subcircuits were then integrated together into an overall circuit design. Second, a 300 W microinverter was presented, followed by a 600 W inverter. The 300 W microinverter is the main inverter used for system within Part II. Finally, a dual-mode inverter was designed. This novel type of inverter can adjust its power rating from 300 to 600 W as needed. Analyses were carried out using key performance indicators for the three inverters.

In this study, the designed inverters are deployed at different scenarios (case studies) to increase the resiliency of an overall system. The different case studies were designed, simulated, and then evaluated based on a novel type of evaluation method created specifically for microinverter systems.

Energy Conservation in Residential, Commercial, and Industrial Facilities, First Edition.
Edited by Hossam A. Gabbar.
© 2018 The Institute of Electrical and Electronics Engineers, Inc. Published 2018 by John Wiley & Sons, Inc.

125

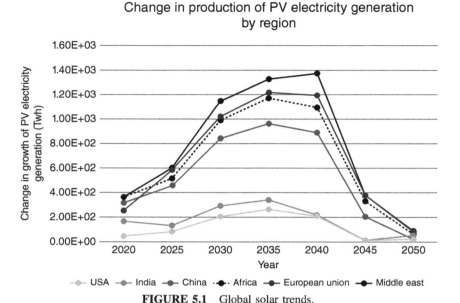

FIGURE 5.1   Global solar trends.

5.1.1 Global PV Trends

As a result of the growing implications of climate change and greenhouse gases, there is growing interest to find clean sources of power generation. Photovoltaic (PV) is the name of a method for converting solar energy from the sun into direct current using semiconductors. Solar energy generation as a whole is an attractive option for combating climate change because it diversifies the energy generation portfolio while providing zero emissions during operation.

Figure 5.1 is shows the projected changes for installed PV generation (TWh) throughout the world [1]. China, India, and the Middle East are expected to have the largest increases of PV. North America, while not having such a drastic increase, is still expected to increase its generation significantly. Overall, the large global increase is largely due to the falling prices of the PV systems and increasing efficiencies. In the mature PV markets, competition is quite competitive; however, legislation and government subsidies are helping push and control the steady increase of PV electricity generation.

5.1.2 Solar PV in Canada

In Canada, solar PV has become one the favored renewable energy technologies due to the social, economic, incentive, and legislative factors. The Canadian federal government promised to try and reach a reduction of emissions to 116 Mt CO_2 equivalent per year in the 2009 Copenhagen Accord [2]. On the provincial

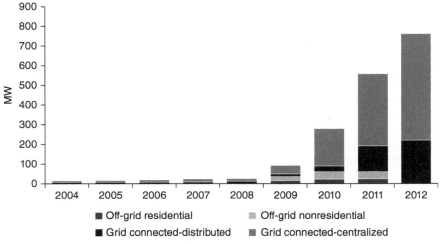

FIGURE 5.2   Solar trend in Canada.

level, the same year (2009) provincial governments in many Canadian provinces established incentives and rebates for PV and other renewable energy sources. This was done to reduce the costs of these systems and allow them to be more economically attractive for people to deploy. These programs have contributed to the rapid growth of PV within the Canadian sector.

Figure 5.2 is from Statistics Canada, which shows the domestic use of photovoltaic technologies in Canada. From 2009, there has been a steady increase in Canada's PV. As stated previously, this is largely a result of provincial programs for PV (and other resources) becoming operational in late 2008/2009 [3]. For example, in 2009 the Ontario provincial government launched the incentive programs feed-in-tariff (FIT) and microfit (micro-FIT) aimed to increase the province's generation of renewable energy. With wind, solar and hydroelectric are the available means of energy generation to qualify for the incentive program. In 2008, the province of Alberta launched regulations allowing Albertans to generate PV and receive credit for power they generate. As a result of these programs becoming operational, PV deployment and capacity have steadily increased, backed by provincial governments.

5.1.3 Problem Statement

A key factor in the long-term success of the photovoltaic industry is the confidence in the reliability of the systems [4]. The three largest expenses of PV systems are typically the inverters, solar panels, and mounts. The prices of inverters have been dropping by about 10% for every doubling of cumulative production in contrast to the PV modules, which have been dropping by 20% cost for the same amount of production [4]. Recently, in The Netherlands it was found that average selling price of solar modules decreased approximately 44.3% in 2012, while that of inverters have decreased only 14% [5]. Despite the dropping costs, inverter's reliability and life span has remained an issue to date as inverters are the most commonly noted causes of system failure triggered in the field [4,6]. These setbacks to system reliability can cause issues and potentially hinder growth of the PV systems.

To date, the PV industry's main focus has been on lowering the price of PV panels and improving the overall efficiency [4]. Given the focus of cost reduction and improved efficiency, the long-term reliability and resiliency of the PV microinverter systems has not been fully developed and has been overlooked. The US Department of Energy (DOE) is currently working with Sandia National Laboratories in the development of accelerated inverter test protocols and standards for reliability. Unfortunately to date, testing of PV inverter reliability is just starting to become more relevant [4]. This leaves a growing need for the development of highly reliable inverters and increased resiliency of PV systems. By enhancing the robustness and performance of PV systems, it will help to aid and support the continued penetration of solar energy into the electricity grid and therefore the overall success of the industry [6].

5.2 BACKGROUND

5.2.1 History of the Inverter

An inverter is a widely applicable device used for a variety of applications. For PV systems, inverters are a fundamental component of the overall system. Inverters are essentially electronic devices that convert direct current (DC) into alternating current (AC) [7]. In PV applications, the inverter is needed to convert the DC generated by solar panels to the AC (whether grid connected or not).

The exact origin of the word inverter cannot be traced with absolute certainty. David Prince is the person who is most likely credited with coining the term [8]. In 1925, Prince published an article in the GE review titled "The Inverter" [8]. In this article, Prince explained that he took the rectifier circuit and inverted it. Taking in DC and outputting AC. He did not physically mean inverting the rectifier devices, but rather he inverted the function or operations of the rectifier [7]. By 1936, Princes' term had spread through the world, and became a common usage term still used today [7].

Originally, rotary converters (called "inverted rotaries") were manufactured until the 1950s, which would transform DC into AC [7]. In the 1950s, semiconductor technology began to emerge, and quickly replaced the inverter rotaries due to their improved characteristics. Today, the IEEE definition of the inverter is "a machine, device or system that changes direct current power into alternating power." As stated earlier, inverters are widely applicable circuits with a broad definition. Therefore, there is a need to further categorize them based on their characteristics. Further history will be discussed in the section on single-phase inverters for residential applications.

5.2.2 Inverter Classification Based on Power Rating

There are many broad classifications for inverters based on a wide variety of things, for example, output power sizes, power stages, control structures, output types, internal components, and so on. Typically, when beginning to study inverters, two main classifications are used as a first introduction: the voltage source inverter (VSI) and the current source inverter (CSI) [9,10]. Practical sources for electronics (batteries, solar panels, etc.) provide either a constant current or a constant voltage. Inverters are typically classified by the source that is supplying the power. A second type of classification that is important for inverters relates to the power output of the device. Typically, power size classification for inverter power sizes are as follows:

$$\text{Large} :> 100\,\text{kW}$$
$$\text{Medium} : \text{from } 10 \text{ to } 100\,\text{kW}$$
$$\text{Small} :< 10\,\text{kW}$$

Microinverters are a subclassification of the small inverter classification and can typically range from 100 to 300 W in output power (this is typically the output power of a single solar panel).

5.2.3 Inverter Market History

5.2.3.1 Evolution of PV Inverters In order to connect the solar panel outing direct current to the grid, a device is needed to transform the power and meet grid standards. Typically, the first generation of inverters for grid connect applications used a single stage (one single power stage). These first generation PV systems are commonly called string, centralized, single-stage systems. More recently, industry has begun shifting toward the use of decentralized inverters with two power stages. A general summary of the evolution of PV systems for residential application along with their configurations are described with reason for the shift [11].

5.2.3.2 Single-Stage Inverters As already stated, the first generation of inverters utilized a centralized single power stage to convert the electricity. This can be seen in Figure 5.3 with string-connected (multiple solar panels connected in series) PV modules.

In order for this system to supply a large enough voltage for the inverter to connect to the grid, the PV panels were arranged in series along with a string diode (protection for panels). The strings (single string shown in Figure 5.3) could then be placed in parallel with one another to reach the desired power level for output.

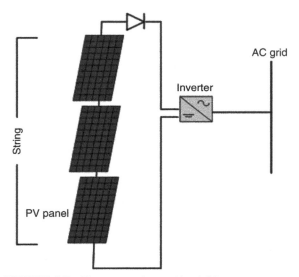

FIGURE 5.3   First-generation residential inverter systems.

While this is a good design, the system has many drawbacks: high-current DC cables were needed, reduced efficiency off overall system due to single panel, reduced efficiency due to degradation factors, efficiency losses due to string diodes, power losses to do one centralized maximum power point tracker (MPPT), and large harmonics [11]. For example, shading on a single solar panel you limit the overall system output and reduce its efficiency. One of the most costly problems of this design was the need to have many PV panels in series to reach the required voltage levels. In order to meet European standards, approximately 16 PV panels needed to be connected in series with each other to form a string (using 45 V per panel to reach 720 Vrms) [11].

5.2.3.3 Two-Stage Inverters In order to overcome some of the shortcomings of the single-stage inverters, a separate design was considered. This is known as a two-stage inverter system and the second generation of PV inverters.

This system is known as two-stage inverter as a first power stage is needed to increase the voltage to the desired level, the second power stage being the inverter. There are a few benefits of changing the overall system design to this one: The efficiency can be improved by optimizing each individual panel, MPPT can be used, shading of a panel or manufacturing defects does not affect the whole string (there is so string), and bulky transformers can be removed, cutting on size and weight, making the inverters smaller and more compact.

Figure 5.4 shows different methods for connecting two-stage inverters. In Figure 5.4a, each of the DC/DC converter stages are connected directly to their own individual inverter stage. In Figure 5.4b, each of the DC/DC power stages are

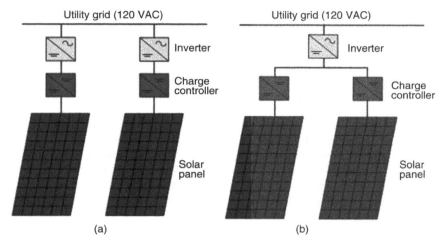

FIGURE 5.4   Second-generation residential inverter systems.

coupled to a single inverter stage. For the purposes of increasing the resiliency for a microinverter system, both parts (a) and (b) are being explored as ways to introduce resiliency into the design of the overall system and will be shown later in this chapter.

5.2.4   Inverter Overview

5.2.4.1   Control Need   Electronic devices often need control systems to control the output, protect the devices, protect internal components, and for many other reasons as well. Inverters are no different and need a variety of control systems to operate, control, and protect the circuit. In addition to protecting and operating the inverter, control circuits are needed for connecting with the grid.

IEEE Std. 1547 provides the overview of the requirements needed to connect a distrusted energy source to the grid. Examples of such requirements are no more the 5% THD, fault clearing/disconnect within 160 ms, and many others.

Therefore, when considering grid-connected applications, a variety of control systems are needed to synchronize voltage and current with the grid for normal operation, ensure correct voltages are outputted, ensure THD is below grid standard, ensure devices are properly disconnected in the event of grid failure, and many other reasons. To further complicate matters, grid fluctuation can be nonlinear due to nonlinear loads being attached/disconnected from the grid. In addition, PV generation is also nonlinear as PV can change its output in nonlinear manners due to weather conditions (temperature, humidity, cloud cover, etc.). Therefore, the control on the inverter is a fundamental and essential part in the overall design.

The general block diagram with a few of the essential control loops for the inverter can be seen in Figure 5.5. The control loops seen in the figure provide and

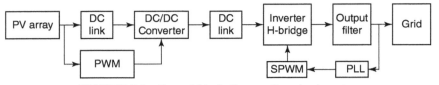

FIGURE 5.5   General block diagram for microinverter.

control the operations for the inverter. In addition, there are many other proposed control loops by researchers, for such things as output current control, filter control, dc link harmonic control, and so on. Traditionally, PI, PID controllers are the most popular controllers used due to their relatively low cost, ease of implementation, and good performance characteristics. In recent years, nonlinear controllers are becoming popular and being integrated into the designs allowing the control systems to become more intelligent over a wide variety of conditions. Nonlinear control typically are improving upon the performance characteristics of the traditional controllers; however, they come with the drawback of requiring higher computation and power draw to do so. Fuzzy logic control for inverters are one of the most common methods for deploying nonlinear control and are becoming more and more popular due to their increasing performance characteristics.

5.2.4.2 *Inverter Control Strategies*   Traditionally, inverters use a pulse width modulation (PWM) as the actuator to adjust output [12]. PWM techniques involve turning switches on and off multiple times in order to create a signal with different pulse widths. This actuates the control system by turning the switches on for longer/shorter periods of time depending on the need of the control systems logic. PWM strategies are not solely limited to inverters and find a wide variety of applications. However, they are widely used for inverter due to their effectiveness and low THD outputs, reduced power losses, and ease of generation and deployment [13].

Sinusoidal Pulse Width Modulation (SPWM) is a technique used by the inverter stage in order to create a sine wave output. PWM is used by the DC converter stage. The two differ from each other in the modulation wave applied. SPWM offers many beneficial characteristics in order to create and adjust output.

In SPWM techniques, two different signals are compared against one another: the modulating reference signal (sinusoidal) and a carrier wave (triangular). The Mosfet gate pulses of the switches are triggered by comparing the two different signals against one another. The width of each pulse is varied in portion to the amplitude of the applied sine wave inputted and can help vary the inverter's output frequency and phase.

Figure 5.6 shows the general graphical representation of the SPWM technique. There are typically two main types of SPWM techniques: unipolar and bipolar triggering. Both have certain advantages and drawbacks versus the other. Bipolar

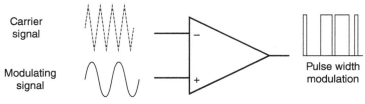

FIGURE 5.6 Inverter control methods.

switching offers a simple method, which is easy to use and implement while offering low distortions and switching losses [14]. Unipolar is more complex than bipolar to implement; however, it has reduced switching losses and generates less EMI [14].

5.2.5 Grid Synchronization

In order for grid-connected inverters to synchronize with the grid, a specialized control system is needed. The most common method used is the phase-locked loop (PLL) control system that synchronizes the inverter output in the electricity grids.

PLL is originally a concept developed in 1932 for radio signals [15]. It was originally published by Appleton in 1923, then by Bellescize in 1932 [16]. PLL quickly became popular and spread to various industries such as communication systems, motor controls, and inverters due to its effectiveness. Essentially, a PLL is a closed loop feedback system that generates an output signal in relation to the frequency and phase of the inputted reference signal.

The control system's main components consist of a voltage-controlled oscillator, a loop filter, and a phase detector. The purpose of the phase detector is to detect the phase difference between the oscillator and the input reference creating an error signal. The error signal is then sent to a loop filter. In order to avoid high transient fluctuations and noise, a loop filter is added to help stabilizes the output from variations. After the loop filter, the output signal is sent to the voltage-controlled oscillator, which adjusts itself accordingly to the input.

The negative feedback forces the error signal to approach zero, at which output and reference signal are in phase with one another [17]. There are many topologies and modifications to the fundamental PLL system shown in Figure 5.7. For example, additional feedback terms are added in order to help increase

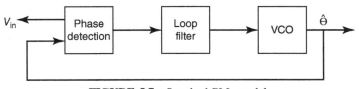

FIGURE 5.7 Standard PLL model.

synchronization speed, improve immunity to noise, and a few other characteristics. Each technique offers its own advantages and disadvantages to the control system. Since PLLs are widely used across a variety of fields, different modifications and techniques are continually being made on PLL systems to improve the performance within that specific field. For inverters, there are many different types and techniques for PLL. Every year, the amount of different PLL methods for three-phase inverters and single-phase inverter is growing. An example of this is the introduction of Kalman filter for single-phase PLL into the overall structure, which reduced the response time [18], or the use of fuzzy logic to reduce overshoot [19]. For single-phase inverter, there are two main types of PLL schemes: stationary frame and synchronous frame [20].

Stationary frame PLLs are a widely used technique for single-phase inverters. The input reference signal for this topology is just the grid voltage (this method does not require additional signals as in others types). This method first employs a sinusoidal multiplier phase detector, then loop filter and VCO [20]. Additional feedback terms have been deployed to this method [20] for improving various performance characteristics such as reduced synchronization speed and its immunity to noise.

Synchronous frame phase lock loop (SRF-PLL) is the second type of common PLL method. Arguably, this is one of the most widely used methods for connecting three-phase systems. This is due to the simple and robust structure that can estimate phase angle and frequency of a three-phase balanced system while having a minimal steady-state error [16,21]. Synchronous frame methods can be utilized within a single-phase system, provided they have an input from an additional orthogonal version to the reference signal. There are two main methods for generating the orthogonal signal for SRF-PLL that exist. First are the methods that generate the orthogonal signal directly from the input signal. Second are the methods that generate the orthogonal signal using the phase angle information and apply an inverse transform to the dq signals. Despite the different methods, it was found that they all converge to the same one point despite their different structures and presentations [22].

For the inverter design discussed in this chapter, a low-complexity PLL method based on mathematical modeling as proposed by Antchev, Pandiev, Petkova, Stoimenov, and Tomova [12,23] is used. This method is based on trigonometric transformations – sine and cosine in a phase detector block. It was chosen because it is easy to implement, provides a quick response times, and is easy to deploy in digital programmable devices.

5.2.6 Key Performance Indicators

5.2.6.1 Introduction Key performance indicators (KPI) are a useful performance measurement tool used in a wide variety of fields. They can be used to evaluate the performance of companies down to single components with circuits. The benefit of KPIs is that they provide a common platform to evaluate the characteristics for the analyzed part in question.

For inverters, KPIs provide a means to gauge the effectiveness of different designs and ensure stakeholder requirements are met, as well as grid standards. An example of this is comparing two different inverter designs against an efficient KPI. Both topologies are compared and contrasted to one another using a common platform. When used in the design stage, the KPI can act as a feedback mechanism to ensure that specific requirements are met. For example, an IEEE Standard states that THD must be less than 5%. During the design stage, the inverter can be tested and if the THD is greater than 5%, the system has to be reiterated upon and design changes made.

KPIs were used as the main evaluation tool for this work herein and used all the way through the design: down from the component-level design all the way to the full design and then even into the full-system-level design.

There are typically four main categories of KPIs: economic KPIs, environmental KPIs, safety KPIs, and performance KPIs. The subcategories for each vary, depending on the application, technology, or even company the KPI criteria are applied to. For this work, only two main KPI categories will be explored: performance and economic.

5.2.6.2 *Performance KPI*   Performance KPIs are used to evaluate the overall effectiveness of the microinverter/inverter.

Peak Operational Efficiency   System efficiency is one of the most important parameters for the design and analysis as it tells the effectiveness of the micro-inverter in converting DC into AC.

$$\eta = \frac{P_{out}}{P_{in}} \cdot 100 \tag{5.1}$$

In the case of the inverter design,

$$\eta = \frac{P_{in} - \sum P_{losses}}{P_{in}} \cdot 100 \tag{5.2}$$

Here P_{in} is the power output from the solar panel, $\sum P_{losses}$ are the losses associated with the boost converter, DC links, inverter-stage filter and will be explored in the design sections.

Peak Operating THD   Total harmonic distortion (THD) is the measurement of the harmonic distortion present within a signal. This is another vital KPI, as IEEE Std. 1547 (Standard for Interconnecting Distributed Resources with Electric Power Systems) [24] requires that the maximum allowable THD from a distributed technology such as solar be less than 5%. As such, no inverter system would even be considered to allow connecting to the distribution grid without having a THD of less than 5%.

$$\text{THD} = \frac{\sqrt{V_{k2}^2 + V_{k3}^3 + V_{k4}^2 + \cdots + V_{kn}^2}}{V_{\text{fund}}} * 100 \tag{5.3}$$

Maximum Synchronization Time   This is a KPI related to the inverter as referred to the maximum amount of time needed for the inverter to synchronize with the grid (reference voltage).

5.2.6.3 *Economic KPI*   Evaluating the economic performance is essential for knowing if the projects are economically feasible or not. It is rare that a project which is not economically feasible is chosen for development over a project that is. As such, assessing some of the economic performance indicators for the design is evaluated to assess the inverter's economic feasibility.

Initial Cost of the Inverter   The economic KPI evaluated is the overall cost of the inverter design. This is strictly for each inverter individually (multiple-inverter designs are present in this chapter) based on the cost of each of the components (shipping costs are not included in the calculation). In order to account for unknowns, as well as a few smaller things (such as wires, etc.), a cost contingency was added to the total cost based on 15% of the total cost of the known components:

$$\text{Initial}_{\text{cost}}(\$) = \sum \text{Parts} \times 0.15 + \sum \text{Parts} \tag{5.4}$$

5.3   INVERTER DESIGN

5.3.1   Circuit Block Overview

This first section provides an overview of the inverter and its subcircuits.

In this section, each of the subcircuit components will be further elaborated, providing the specific circuit topology used, the governing equations, and the PSIM simulation model. Figure 5.8 was provided first, in order to have an overall idea of the entire system and how the subcircuits are fitted into the overall design. Component values for the design were first found using the governing equations. The second step consists of a PSIM model for the specific subcircuit based on the calculated values. From there, a search was done to find components readily available on the marketplace, with values close to the design calculations. Sometimes iterations are needed to be done in order to adjust the calculated values to find available components. In addition, in order to simulate an inverter as realistic a way as possible, the internal resistances of all components were put into the overall simulation. Sometimes this included the addition of extra resistors into the circuit design, simulating the internal resistance of inductors, capacitors, and so on (see Figure 5.9).

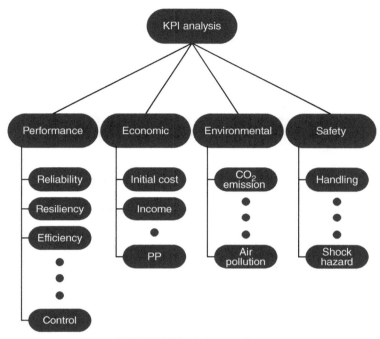

FIGURE 5.8 KPI overview.

5.3.2 Solar Panel Used

First, starting with the solar panel, a 300 W solar panel was chosen from a local manufacture. This was done in order to utilize local components, save on shipping costs, and ease in customer service. This company is Ecilpsall located in

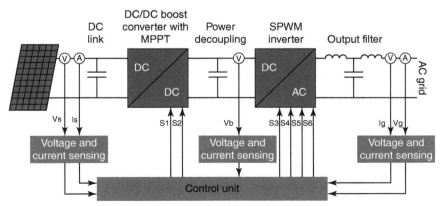

FIGURE 5.9 Inverter overview.

Electrical data

Type	NRG72 300 M	NRG72 305 M	NRG72 310 M	NRG72 315 M	NRG72 320 M	NRG72 325 M
Rated maximum power at STC-Pmax(W)	300 W	305 W	310 W	315 W	320 W	325 W
Maximum power voltage – Vmp(V)	36.54	36.69	36.78	36.88	36.97	37.22
Maximum power current – Imp(A)	8.22	8.32	8.43	8.51	8.66	8.73
Open circuit voltage – Voc(V)	44.87	44.99	45.12	45.28	45.39	45.60
Short circuit current – Isc(A)	8.73	8.84	8.92	8.95	8.97	8.98
Operating temperature	−40 to +85°C					
Max system voltage	1000C (IEC) / 600V (UL)					
Fuse rating	15A					
Power tolerance	0 to 5W					
Temperature Coefficients	Pmax	−0.438%/°C				
	Voc	−0.341%/°C				
	Isc	0.047%/°C				

FIGURE 5.10   Solar panel data.

Scarborough Ontario. From their local product line, a 300 W solar panel, the NRG72 300 M, was chosen.

As shown in Figure 5.10, the rating and data for the NRG72 300 M panel were taken and provided the first step in the design. The maximum voltage, or nominal voltage, of the solar panel is 36.54 V [25]. These are the crucial data needed for the design of the first power stage and the maximum power point tracking. In addition, the maximum current data are taken, along with the short circuit current (maximum output current), and the limits for the circuit provided. In order to properly size components on the market and provide a safe margin for operating conditions, the components were oversized from maximum values. For example, the open-circuit voltage for the solar panel is 44.87 V [25], which can be seen in the data sheet provided in Figure 5.10. Using these data, the first DC link capacitor was then sized with a sufficiently large voltage rating higher than 45 V in order to ensure it can operate safely with the design parameters for the circuit.

5.3.3   DC–DC Converter Subcircuit Design

5.3.3.1 Boost Converter   A boost converter is a DC–DC converter that increases the output voltage from the source, and then applies it to the load. This is the first power stage of the inverter circuit. Each boost converter typically contains at least two semiconductors and at least one energy storage element (typically have two or more).

5.3.3.2 *Governing Equations* The following section outlines the governing equations for a boost converter operating in continuous conduction mode [15].

At charging interval,

$$i_L(t) = \frac{1}{L}V_{in}t + I_L(0), \quad 0 \le t \le DT \tag{5.5}$$

where $I_L(0)$ = initial inductor curent at $t = 0$, V_{in} = voltage across inductor, $i_L(t)$ = current through inductor.

When switch is turned off at $t = DT$, the inductor voltage becomes

$$i_L(t) = \frac{1}{L}(V_{in} - V_0)(t - DT) + I_L(DT) \tag{5.6}$$

$$DT = t < T$$

Evaluating (5.5) and (5.6) such that during steady-state operation, the net change of current between the on/off state must equal zero:

$$\frac{1}{L}V_{in}\,DT - \frac{1}{L}(V_{in} - V_0)(1 - D)T = 0 \tag{5.7}$$

Solving we get,

$$\frac{V_0}{V_{in}} = \frac{1}{1 - D} \tag{5.8}$$

There is minimum critical value of inductance such that anything above would be in continuous conduction mode and anything below would be in discontinuous conduction mode.

$$L_{min} = \frac{RT(1 - D)^2 D}{2} \tag{5.9}$$

Output ripple for the boost converter is given by the following equation:

$$\frac{\Delta V_0}{V_0} = \frac{D}{RCf} \tag{5.10}$$

5.3.3.3 *PSIM Model* The following shows the boost convert circuit used in the subcircuit drawing of the figure in Section 5.3.1.

The left-hand side of Figure 5.11 is the low-voltage side of the circuit or the solar panel input. The right-hand side is the high-voltage portion, or boost converter output. R_L1 is included to model the internal resistance of the inductor,

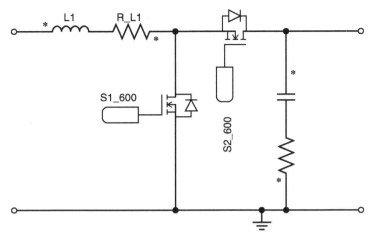

FIGURE 5.11   PSIM boost converter.

taken from the data sheet. In addition, the capacitor also has a resistor added in order to model the ESR for the capacitor. For the Mosfets, MPSIM model has the ability to input the internal resistance; therefore, external resistances were not required to be added to them. The values for the components will be seen in the bill of materials for the inverter in Section 5.3.13.

5.3.4   DC Link

5.3.4.1   Purpose   There are of two different DC link capacitors within the microinverter design: first, connecting the solar panel to the boost converter; second, connecting the boost converter to the H-bridge. This component is used an intermediate device between power system input and output. Effectively, they are there to help maintain a constant supply. For VSI, capacitors are used, and for CSI, inductors are typically used.

5.3.4.2   Sizing   As a result of the pulsating output power of the inverter (downstream), an AC ripple at twice the output frequency occurs along the DC link before the inverter power stage (upstream). This is known as second-order harmonics (SHC). These second-order harmonics will penetrate into the front end of the DC–DC boost converter and can reduce the overall efficiency unless otherwise managed.

Assume the grid current and voltage are

$$v_g(t) = \widehat{V}_g \cos(\omega_g t) \tag{5.11}$$

$$i_g(t) = \hat{I}_g \cos(\omega_g t - \phi) \tag{5.12}$$

Instantaneous power output from the grid is therefore

$$P_{out}(t)\widehat{V}_g I_g \cos(\omega_g t)\widehat{\cos}\ (\omega_g t - \phi) = V_g^{rms} I_g^{rms}\cos\phi + V_g^{rms}\ I_g^{rms}\cos(2\omega_g t - \phi) \tag{5.13}$$

Written another way

$$P_{out}(t) = S\cos\phi + S\cos(2\omega_g t - \phi) \tag{5.14}$$

where S is the apparent power (VA). Assume (I) instantaneous power output from the inverter is equal to the instantaneous power input to the inverter. (II) DC link voltage on the input has a nominal voltage of V_{dc}. Therefore,

$$P_{in}(t) \cong P_{out}(t) \tag{5.15}$$

$$V_{dc} \cdot i_{dc}(t) = S\cos\phi + S\cos(2\omega_g t - \phi) \tag{5.16}$$

Now assume the DC capacitance filters out the high switching frequency components in the DC current $i_{dc}(t)$, which can be separated into the DC component I_{dc} and an AC component $i_{dc,ripple}(t)$:

$$V_{dc}{}^* i_{dc,ripple}(t) = S\cos(2\omega_g t - \phi) \tag{5.17}$$

Rearranging yields

$$i_{dc,ripple}(t) = \frac{S}{V_{dc}}\cos(2\omega_\gamma t - \phi) = \hat{I}_{dc,ripple}\cos(2\omega_g t - \phi) \tag{5.18}$$

The DC link is used to connect one part of the electric circuit to another. That is, the first stage to the second stage and solar to the first stage. They are meant to reduce noise and distortions in the circuit caused by various elements. Normally electrolytic capacitors are used because of their significantly lower costs when compared to other capacitors (film, aluminum, etc.). Unfortunately though, using electrolytic capacitors has also been one of the main reasons for inverter failures in recent years because of their shortened life span under fluctuating temperatures that the microinverters are exposed to.

Inverter warranties range from 3 to 5 years, while PV come with a warranty of approximately 20 years [4]. There is little information that has been published regarding the failure modes of the inverters; however, the US DOE at a workshop for solar power agreed upon the fact that most urgent problem facing the inverters is the DC-bus capacitor linking [4].

An alternative to electrolytic capacitors is film capacitors. These provide significantly better characteristics under fluctuating temperatures. The trade-off, however, is that they are significantly more expensive (can easily be four times larger) than electrolytic capacitors of an equivalent Farad. Therefore, if using film capacitors, a balance needs to be struck, as the capacitor has a sufficiently low Farad value, to keep the cost low. However, it should also have a high enough Farad value to help in the reduction of the second-order harmonics. As such Chen proposed the following equation for the size of the decoupling capacitor [26]:

$$C = \frac{S}{2 \cdot \pi \cdot f \cdot V_{dc} \cdot \Delta V_{DC}} \tag{5.19}$$

5.3.5 Inverter Topology Subcircuit Design

5.3.5.1 *PSIM Model*   For this inverter power stage, a full bridge schematic was chosen in order to simplify the design of the microinverter.

The two wires on the far left-hand side of Figure 5.12 are the connections to the positive and negative DC rails of the high-voltage DC. The two wires on the right-hand side are the output of the inverter stage. Finally, the connections to the SPWM subcircuits can be seen connected to the gates of the Mosfets.

5.3.6 SPWM Design

5.3.6.1 *Overview*   In order to trigger the Mosfets for an H-bridge configuration, a method is needed. By varying the duty cycle of the pulses being applied to

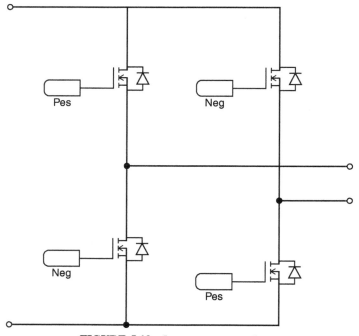

FIGURE 5.12   Inverter stage topology.

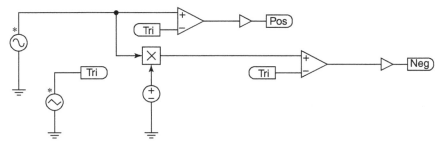

FIGURE 5.13   SPWM inverter firing circuit.

trigger the Mosfets (and therefore the Mosfets), a pure sine wave output can be obtained with the addition of a filter. Digital firing was desired as it can minimize hardware circuitry, increase resolution and, therefore, minimize distortions [27]. In order to generate this signal, SPWM switching was used. In order for it to operate, upper and lower switches in the same inverter leg (S1 and S4, then S2 and S3) are turned on, while the other is turned off. Because of this, only two reference signals were chosen: a reference modulating sine wave v_m (to be link with the PLL and discussed in Section 5.3.10) and a triangular carrier wave v_{cr}.

5.3.6.2 PSIM Model   The firing circuits for inverters vary depending on the semiconductors used as well as the desired output. As stated earlier, for the inverter a sine pulse width modulation circuit was chosen, which can be seen in Figure 5.13.

Two signals are generated and compared to one another. The first is a sine wave genertor as represented by the source of the far left-hand side in the figure. This source will be removed later and integrated with the PLL later. The second source is a triangluar wave generator. The triangluar wave is operating with a peak-to peak voltage of 10 V and 20 kHz. The sine wave generator is operating at the desired 60 Hz. The output of these two signals are sent to the comparator and trigered for every positive/negative cycle, respectivtly. This effectivly tirggers the gates of the inverters in switches Q1 and Q3, allowing current to flow in one direction of the inverter. In order for current to flow in the other direction of the inverter, Q2 and Q4 gates need to be triggered. This is accomplished by inverting the sine wave and comparing with a second comparator cicuit.

5.3.7 Filter Subcircuit Design

5.3.7.1 Inverter Need   The output filter is a fundamental component of the inverter design. The filter is necessary in order to connect to the grid as it filters out the high-frequency SPWM signal and converts it into a sine wave. This is necessary in order to meet IEEE 1547 standard, which states that inverters must have a THD of less than 5% [24]. There are many different types of filters

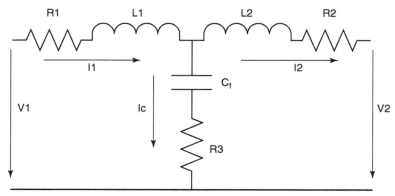

FIGURE 5.14   General drawing LCL filter.

available: L, LC LCL, and LLCL. For the purposes of this design, an LCL filter was chosen over other designs as it provides better attenuation and low ripple current distortions [28] and contained high-efficiency values [29].

5.3.7.2 *Governing Equations*   The following section outlines the LCL filters governing equations and process to iterate to the correct value.

Transfer Function   Figure 5.14 shows the filter model of the LCL filter with the addition of the internal resistances. V1 on the left-hand side refers to the side connected to the inverter H-bridge, V2 on the right-hand side refers to the output connected to the grid. The transfer function derived from the circuit is

$$G(s) = \frac{i_2(s)}{v_1(s)} = \frac{R_3 C_f s + 1}{L_1 L_2 C_f s^3 + (L_1 + L_2) R_3 C_f s^2 + (L_1 + L_2)s} \tag{5.20}$$

Equations   The following section outlines the equations used to calculate the values for the filter as proposed by Refs. [28,30]:

The resonant frequency is

$$\omega_{\text{res}} = \sqrt{\frac{L_1 + L_2}{L_1 L_2 C_f}} \tag{5.21}$$

Total filter inductance calculation is given by

$$L = \frac{V_{\text{dc}}}{4\, I_{\text{rated}} \Delta_{\text{ripple}} f_s} (1 - m_a) m_a \tag{5.22}$$

where the rated utility current, Δ_{ripple}, is the maximum ripple percentage (5–25%), V_{dc} = voltage DC link, f_{s} is the switching frequency, and m_{a} is the modulation index [28].

By adding a second inductor, the inductor is divided into two parts based on the following equation:

$$L_1 = aL_2 \tag{5.23}$$

where a ($a \geq 1$) is the inductance index calculated using the switching harmonic current attenuation ratio:

$$\sigma = \frac{i_2}{i_1} = \frac{(1+\alpha)r}{\alpha(1-r)-r} \tag{5.24}$$

where

$$r = \frac{1}{L_2 C_f \omega_{\text{sw}}^2} \tag{5.25}$$

In addition,

$$\frac{r}{1+\alpha} = \frac{\Delta_{\text{ripple}} V_{\text{rated}} \omega_0}{2\pi^2 V_{\text{dc}} f_{\text{sw}} \alpha} \tag{5.26}$$

where Δ_{ripple} is maximum ripple magnitude percentage (5%–25%) [28], V_{rated} is the rated utility voltage, a is the reactive power factor, and ω_0 is the utility frequency.

Filter capacitance can be found through the following equation:

$$C_f = \frac{Q_{\text{re}}}{\omega_0 V_{\text{rated}}^2} = \frac{\alpha P_{\text{rated}}}{\omega_0 V_{\text{rated}}^2} \tag{5.27}$$

Finally, the damping resistor can be found from the following equation:

$$R_3 = \frac{1}{3\omega_{\text{res}} C_f} \tag{5.28}$$

Procedure   The following is the iterative procedure for LCL filter design selection based on the equations above.

Using Equations 5.21–5.28, the following procedure is outlined in Ref. [28] as a method to first calculate filter component values and then to iterate upon them until they converge onto the appropriate values for the filter. Figure 5.15 shows the process proposed and it was used for the selection of the filter components for both inverter designs.

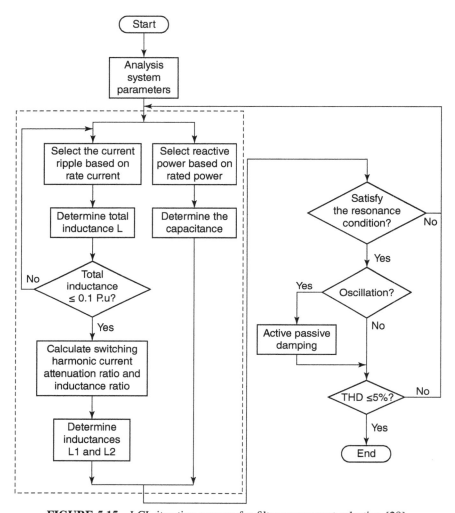

FIGURE 5.15 LCL iterative process for filter component selection [28].

5.3.7.3 *PSIM Model*

The Figure 5.16 shows the PSIM model created for the inverters. As with other models created, the internal resistances were included in order to simulate as realistic a model as possible.

The filter in Figure 5.15 was chosen and designed such that a 60 Hz signal can pass through and effectively remove the higher frequencies within the output signal, thus reducing the unwanted harmonics.

Figure 5.17 shows the frequency analysis of the output filter for the 300 W scenario. It can be seen that the voltage (bottom graph) passes at 60 Hz and begins to filter out the higher frequency values. The current (top graph) is similar with the

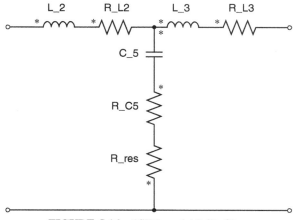

FIGURE 5.16 PSIM model LCL filter.

same passing frequency. This filter helps smooth out and turn the SPWM into a sinusoidal signal and help reduce harmonics for the outputting alternating current. It also follows the same trends as shown in by Ref. [28].

5.3.8 Maximum Power Point Tracking Control Loop Design

5.3.8.1 Purpose A maximum power point tracker is a term used to describe techniques to maximize the solar output from the panel. While MPPT techniques are widely used for photovoltaics, they also have other applications with such thing such as optical power transmission and wind turbines. The need for MPPT arises as

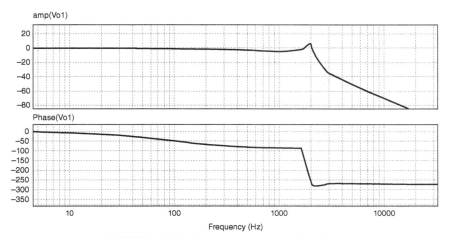

FIGURE 5.17 Frequency analysis LCL filter.

solar cell V-I characteristics are nonlinear in natural and can fluctuate based upon irradiance, outside temperature, cloud cover, and other conditions. Despite this nonlinearity, there is a specific point (maximum power point or MPP) for the solar panel at which the maximum power can be extracted. MPPT then is a technique used in order to find and locate the MPP during the various conditions faced by the panel.

5.3.8.2 Choice in Method   There are many different techniques and methods for MPPT. For the purposes of this work, the P&O method was utilized. The reason this method was chosen is, first, a preexisting model comes with PSIM, therefore, making it is easy to implement in the software simulation. The second and more important reason for choosing P&O method is that when compared to the other popular methods (constant voltage method, incremental method, short circuit method, and open voltage), it was found to generate the highest amount of energy [31]. Finally, the P&O method is easy to implement on hardware micro-controller and can easily be purchased off the shelf if needed [32]. Taking all this into account, the P&O method was deemed the preferred method to be used.

5.3.8.3 PSIM Model   The following section shows the preexisting MPPT P&O method available within PSIM that was integrated into the overall designed inverters (see Figure 5.18).

The PSIM model takes in the voltage and current values from the solar panel, then takes the derivative of these and compares the results to zero. If greater than zero, a trigger closes the dU1 switch, thereby increasing the voltage reference. If less than zero, then the inverse dU2 switch is triggered and the reference decreases in voltage. This voltage is then compared to the voltage generated by solar panel,

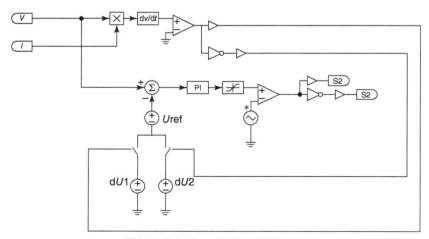

FIGURE 5.18   PSIM MPPT model.

creating an error signal between the voltage and voltage reference. This is then sent to a PI controller and comparator to trigger the boost converter accordingly.

5.3.9 Grid Synchronization – PLL Control Design

5.3.9.1 Overview As stated previously, PLL systems are a critical control system for grid-connected microinverters in order to synchronize to the grid. This chapter uses a PLL method based on trigonometric transformations as proposed by Ref. [12,23] and integrated into the overall design for the inverter designed here in. The advantage of this PLL is its insensibility to changes of amplitude input signal after synchronization has occurred, reduced settling time for frequency changes, reduced steeling time for phase changes, and finally ease of implementation into microcontrollers [12,23].

5.3.9.2 Operational Theory As already stated, one of the major advantageous of this proposed PLL is its insensibility to changes in the input after synchronization has occurred. This is because most often disturbances have a short duration and usually consist of incrementing or dementing the voltage.

The assumptions for this method are that

$$v_{in} = U_M \sin v$$
$$v_{sync} = \sin \hat{v}$$

With this in mind, the operation of the proposed PLL is based on the following mathematical equations [12,23]:

$$U_M \sin v.\cos \hat{v} = U_M \frac{1}{2[\sin[(v - \hat{v}) + \sin(v + \hat{v})]]} \tag{5.29}$$

$$-U_M \cos v.\sin \hat{v} = -U_M \frac{1}{2[\sin[(v - \hat{v}) + \sin(v + \hat{v})]]} \tag{5.30}$$

After summing Equations 3.26 and 3.27, the basic trigonometric relations can be obtained:

$$U_M \sin v.\cos \hat{v} - U_M \cos v \sin \hat{v} = U_M \sin(v - \hat{v}) \tag{5.31}$$

The equation is incorporated into the standard PLL model as shown in Figure 5.19.

The operating principle is such that the right-hand side of Equation 5.29 is used as an error signal for the closed loop PLL. This signal is then sent to the PI controller in PSIM, which then outputs its values to the input of the VCO. This process continues to adjust until the phase difference and frequency between the input and output signal goes to zero. At zero, the PI controller is equal to zero and the steady-state operation has been reached.

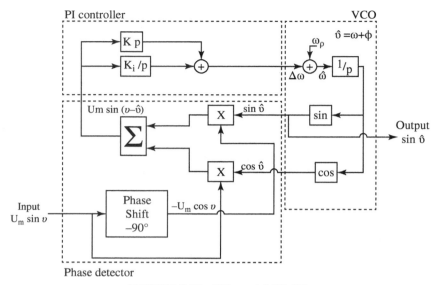

FIGURE 5.19 PLL model [12,23].

5.3.9.3 PSIM Model

In order to simulate the PLL, a model was created in PSIM according to the equations and method proposed in Refs [12,23]. The simulation model contains summer block, PI block, multipliers, sine wave, cosine wave, and a time delay block. The time delay is acting as a phase shifter and generates cosine wave feeding into cosine multiplier. The multiplier block is the right-hand side of Equation 5.29, which then sends the output values into the PI controller block. This block preforms the operation of being the PI regulator in the PLL. This outputs the controlled signal to the VCO portion, which in turn outputs a sine wave. The output from the VCO is then fed back into the phase detector and the process begins again. In addition, the VCO output is fed into the SPWM as the reference sine wave signal to trigger the inverters (see Figure 5.20).

5.3.10 300 W PSIM Circuit Design

Figure 5.21 shows the overall circuit design for the 300 W inverter in PSIM based on earlier subcircuit designs.

Starting from left to right, it can be seen that there is the solar panel (with the specifications of NRG72M within the model). Second is the DC input filter, followed by the boost circuit as shown in Figure 5.21. Next is the DC link capacitor used to couple the boost converter to the inverter. Finally, this is followed by the second power stage, or the H-bridge inverter. Finally, the output is connected to the LCL filter, which outputs to the grid. In the control circuit at the bottom of the Figure 5.21, the subcircuits PLL and MPPT are contained within.

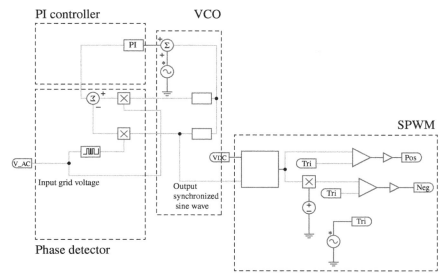

FIGURE 5.20 PSIM model integrated with SPWM.

5.3.10.1 *Bill of Material 300 W Design* Table 5.1 shows the components used, values, cost, and amount of parts for the 300 W microinverter design.

Parts and their respective prices were obtained from the Digikey and Mouser Web sites. Unit prices (highest value) were taken in order to provide a conservative estimate.

5.3.11 600 W Inverter Circuit Design

The following section outlines the simulation results for the 600 W PSIM circuit design.

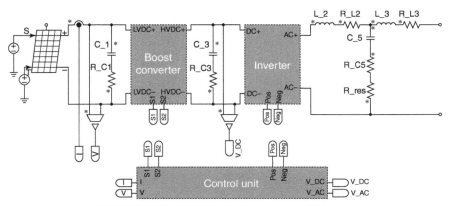

FIGURE 5.21 300 W microinverter overall design.

TABLE 5.1   BOM 300 W Microinverter [33,34]

Component Description	Component Values	Manufacturer	Amount	Cost Per Unit ($CAN)
DC link capacitor	4.7 μF	Panasonic ECA2CM4R7	1	0.49
Boost inductor	9 mH	Wurth Elektronik	1	34.98
Boost capacitor	10 mF	Electronicon E50.R29-505NTO	2	70.65
N-channel Mosfets	N/A	Sanken SKP253VR	6	2.425
DC link capacitor	5 mF	Electronicon E50.R29-505NTO	1	70.65
Filter inductor	390 μH	J.W.Miller 1140-391K-RC	1	10.82
Filter inductor	100 mH	Hammond 195T5	1	135.44
Filter capacitor	18.3 μF	TDK B32794D3205	1	1.45
Filter capacitor	18.3 μF	B32794D2156	1	6.67
Filter resistance	2 Ω	Bourns PWR263S-20	1	3.51
MPPT + PLL chip	N/A	TI C2000 Piccolo	1	16.43
Boost	N/A	Allegro ACS712 ELCTR-20A	1	2.69
PLL grid	N/A	Honeywell CSLA2CD	1	37.03
			Total	476.01
Extra (PCB, wires, resistors, heat sink, unknown, etc.)				+15%
			Estimated total	547.42

The 600 W inverter circuit design follows the same topology as the 300 W microinverter with the addition of a second solar panel parallel to the input. The internal subcircuit components are all different and represent the 600 W values found (see Figure 5.22). The list of components for the 600 W inverter can be seen in Table 5.2.

5.3.11.1   600 W Inverter Design   Table 5.2 shows the values used in order to simulate the 600 W inverter designed.

As with the 300 W inverter, parts and their respective prices were obtained from the Digikey and Mouser Web sites and unit prices were used.

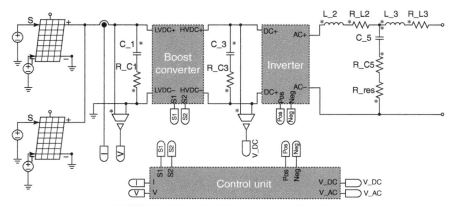

FIGURE 5.22   600 W inverter overall design.

TABLE 5.2   BOM 600 W Inverter [33,34]

Component Description	Component Values	Manufacturer	Amount	Cost Per Unit ($CAN)
DC link capacitor	4.7 µF	Nichicon UDB1H4R7MPM1TD	1	0.39
Boost inductor	30 mH	Hammond 195p20-ND	1	453.37
Boost capacitor	3.3 mF	TDK B43456A9338M	1	110.26
N-channel Mosfets	N/A	Sanken SKP253VR	6	2.425
DC link capacitor	3.3 mF	TDK B43456A9338M	1	110.26
Filter inductor	1 mH	Schurter Inc.	1	33.61
Filter inductor	50 mH	Hammond 195R19	1	267.38
Filter capacitor	37 µF	TDK (22 µF) B4356A2229M	1	213.12
Filter capacitor	37 µF	TDK B4354A2159M (15 µF)	1	191.03
Filter resistance	2 Ω	Bourns PWR263S-20	1	3.51
MPPT + PLL chip	N/A	TI C2000 Piccolo	1	16.43
Boost	N/A	Allegro ACS712 ELCTR-20A	1	2.69
PLL grid	N/A	Honeywell CSLA2CD	1	37.03
			Total	1453.7
Extra (PCB, wires, resistors, heat sink, unknown, etc.)				+15%
			Estimated total	1671.7

5.3.12   Dual-Mode Inverter Design

The following section provides an overview of a novel inverter design. Effectively, this inverter has two modes of operation it can achieve: first, a 300 W mode for normal operation, followed by a 600 W mode for operation under fault.

The need for this dual-mode inverter arose when assessing the resiliency strategies for the microinverter system.

One of the proposed methods required that upon fault in an inverter, a paired inverter would modify its internal structure in order to compensate for the failed inverter. The solar panel attached to the faulted microinverter would be reconnected to the operation microinverter redirecting its power flow through the functioning microinverter. The function microinverter would modify its internal components (to 600 W) in order to maintain system output. A more complete explanation will be discussed in Section 5.8.

5.3.12.1 DC/DC Converter   For the microinverter to modify its output power level, the subcircuit should modify and adjust its internal components accordingly.

Figure 5.23 shows how the boost converter was modified to handle two power output sizes. Both the inductors of the 300 and 600 W inverters are placed in parallel to one another, with relays on both sides of them. In addition, the capacitors of both inverters are placed in parallel and connected/disconnected to the circuit with a relay. It should be noted that the extra resistances corresponding to the relays' internal resistance were added. Under normal operation, the relay for the 300 W inverter is closed, while the 600 W inductor and capacitor relays are opened. In the event of a fault (in the other microinverter), the switches from the 300 W boost converter open,

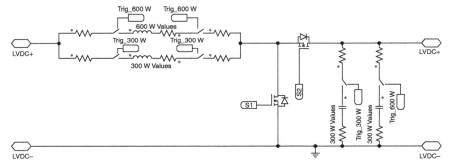

FIGURE 5.23 PSIM boost converter for dual-mode inverter.

and the switches for the 600 W boost converter close. This effectively allows the inverter to reconfigure the boost converter in order to handle the various output powers. It should be noted that the N-Channel Mosfets do not need modification, as both modes are well within safe operating limits of the Mosfets.

5.3.12.2 H-Bridge Inverter No changes were needed for the second power stage (H-bridge) and the internal subcircuit design is the same as in Section 3.5.1. This is for the same reason as the boost converter Mosfets, in that they can safely operate in both modes. In addition, they are low-cost and low-internal resistances from a large-scale and reliable manufacturer.

5.3.12.3 Filter Design Similar to the boost converter, modifications were needed for the filter design to accommodate the various power levels.

Figure 5.24 shows how the modifications were made to the filter. Inputs of the filter are shown on the left-hand side and coupled to the H-bridge. The right-hand side shows the output, which connects to the grid. Similar to the boost converter,

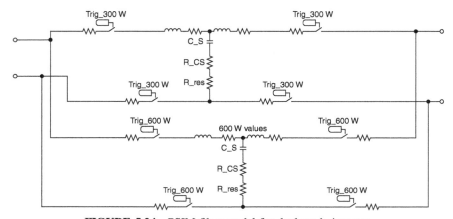

FIGURE 5.24 PSIM filter model for dual-mode inverter.

TABLE 5.3   BOM Dual-Mode Inverter [33,34]

Component Description	Component Values	Manufacturer	Amount	Cost Per Unit ($CAN)
		300 W microinverter		
DC link capacitor	4.7 μF	Panasonic ECA2CM4R7	1	0.49
Boost inductor	9 mH	Wurth Elektronik	1	34.98
Boost capacitor	10 mF	Electronicon E50.R29-505NTO	2	70.65
N-channel Mosfets	N/A	Sanken SKP253VR	6	2.425
DC link capacitor	5 mF	Electronicon E50.R29-505NTO	1	70.65
Filter inductor	390 μH	J.W.Miller 1140-391K-RC	1	10.82
Filter inductor	100 mH	Hammond 195T5	1	135.44
Filter capacitor	18.3 μF	TDK B32794D3205 (3.3 μF)	1	1.45
Filter capacitor	18.3 μF	B32794D2156 (15 μF)	1	6.67
Filter resistance	2 Ω	Bourns PWR263S-20	1	3.51
MPPT + PLL chip	N/A	TI C2000 Piccolo	1	16.43
Boost	N/A	Allegro ACS712 ELCTR-20A	1	2.69
PLL grid	N/A	Honeywell CSLA2CD	1	37.03
		600 W inverter		
Boost inductor	30 mH	Hammond 195p20-ND	1	453.37
Boost capacitor	3.3 mF	TDK B43456A9338M	1	110.26
Filter inductor	1 mH	Schurter Inc.	1	33.61
Filter inductor	50 mH	Hammond 195R19	1	267.38
Filter capacitor	37 μF	TDK (22 μF) B4356A2229M	1	213.12
Filter capacitor	37 μF	TDK B4354A2159M (15 μF)	1	191.03
Filter resistance	2 Ω	Bourns PWR263S-20	1	3.51
Relays: boost, filter	N/A	TE PB1014-ND	14	6.19
			Total	1834.95
Extra (PCB, wires, resistors, heat sink, etc.)				15%
			Estimated total	2110.2

the filter uses relays in order to connect the appropriate filter for the desired power level.

5.3.12.4   *Dual-Mode Inverter*   Table 5.3 shows the values used to simulate the dual-mode inverter designed. The parts needed for this inverter are effectively the addition of both the 300 and 600 W inverter components together.

However, additional Mosfets are not needed, and the Mosfets from one inverter will suffice for the two. The extra component needed is just extra relay. The prices reflect the components taken from the Digikey Web site and Mouser Web site for unit price values.

5.4   SIMULATION RESULTS

Power electronics simulation software (PSIM) was used to model the three inverters proposed.

5.4.1 300 W Microinverter

Figure 5.25a shows the output of the solar panel or input into the inverter system. Figure 5.25b shows the input voltage and the input current. From the figure, it can be seen that the MPPT is working by keeping the voltage around 36 V and current 8 A. The boost capacitor has approximately 4 W power switching losses in addition to the resistive losses. Figure 5.25b shows the voltage output after the first stage and across the DC link capacitor. This pulsation in the voltage is due to the second-order harmonics generated within the circuit. However, to minimize the harmonics, a little slightly larger capacitor was used. The fluctuations can be seen to be around +/−0.3 V.

Figure 5.26a shows the output results of the 300 W microinverter. The top-most graph of Figure 5.26a corresponds to the output voltage, middle graph shows the output current, and the bottom-most graph shows the grid reference voltage. From

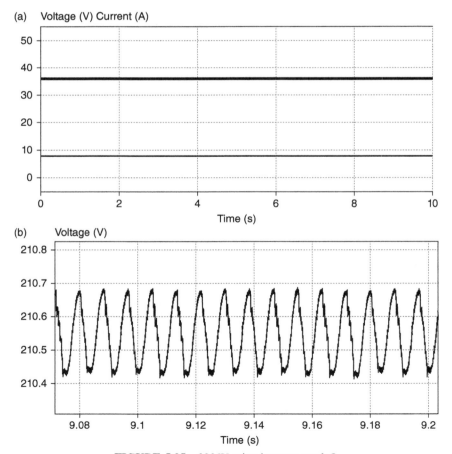

FIGURE 5.25   300 W microinverter result I.

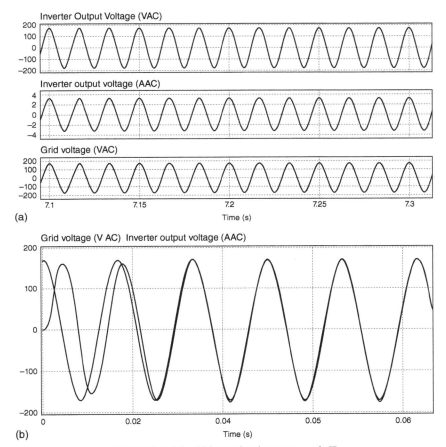

FIGURE 5.26 300 W microinverter result II.

the figures, it can be seen that the system is working in unison, outputting 120.1 V (rms) and 2.27 A (rms). Further analysis will be shown in the Table 5.4 in Section 5.4.4. Figure 5.26b shows the synchronization of the microinverter with the grid. A 60° phase shift was initially put between the microinverter and the grid. The figure shows the synchronization from the microinverter (blue) to the grid voltage (red), which takes approximately 35 ms to resynchronize with the PLL method used. A further analysis will also be shown in Table 5.4.

5.4.2 600 W Inverter

Figure 5.27a shows the steady-state operation of the solar panels (two 300 W) power into the inverter. The red line shows the voltage input maintained at 36 V, while the blue line shows the current approximately at 16 A. The power switching

TABLE 5.4   KPI Analysis Inverter

Key Performance Indicator	300 W Microinverter Results	600 W Inverter Results	Dual-Mode (300 W/600 W) Inverter Results
Operating efficiency	$P_{in} = 287.44$ W $P_{out} = 274.44$ W $\eta = 95.47\%$	$P_{in} = 574.2$ W $P_{out} = 508.2$ W $\eta = 88.51\%$	300 W mode $P_{in} = 289.1$ W $P_{out} = 259.6$ W $\eta = 89.8\%$ 600 W mode $P_{in} = 574.62$ W $P_{out} = 489.3$ W $\eta = 85.17\%$
THD	2.11E-002	1.735E-002	300 W mode 2.54E-002 600 W Mode- 1.37E-002
PLL synchronization response time	Phase shift $0° = 21.55$ ms $45° = 26.81$ ms $90° = 32.29$ ms $135° = 47.37$ ms $180° = 34.62$ ms $225° = 32.36$ ms $270° = 22.07$ ms $315° = 18.13$ ms $360° = 21.41$ ms	Phase shift $0° = 13.93$ ms $45° = 28.93$ ms $90° = 26.84$ ms $135° = 41.13$ ms $180° = 35.12$ ms $225° = 24.48$ ms $270° = 21.83$ ms $315° = 12.17$ ms $360° = 14.27$ ms	Phase shift $0° = 14.17$ ms $45° = 20.65$ ms $90° = 26.35$ ms $135° = 41.67$ ms $180° = 34.47$ ms $225° = 23.47$ ms $270° = 22.14$ ms $315° = 11.77$ ms $360° = 14.19$ ms
Initial cost	\$547.4115 (CAN)	\$1671.8 CAN)	\$2110.2(CAN)

losses from the boost converter are significantly higher than in the 600 W with about 19 W in power loss contributing to the overall power losses within the circuit design. Figure 5.27b shows the output from the first power stage. The second-order harmonics across the DC link are fluctuating approximately around +/−1.2 V.

Figure 5.28a shows the output from the 600 W inverter. The top-most graph of Figure 5.28a shows the output voltage. The middle graph shows the output current, and the bottom graph shows the grid voltage. It can be seen that the synchronization has taken place and the system is working in unison with an output voltage 119.6 V and current 4.22 A (rms). Figure 5.28b shows the grid synchronization with a phase shift of 60° from the grid voltage to the system. Synchronization occurs in approximately 40 ms. Further analysis of this inverter will be shown in Table 5.4.

5.4.3   Dual-Mode Inverter

5.4.3.1 Normal Operation (300 W Mode)   Figure 5.29a shows the steady-state operation of the microinverter operating under 300 W mode. One single solar

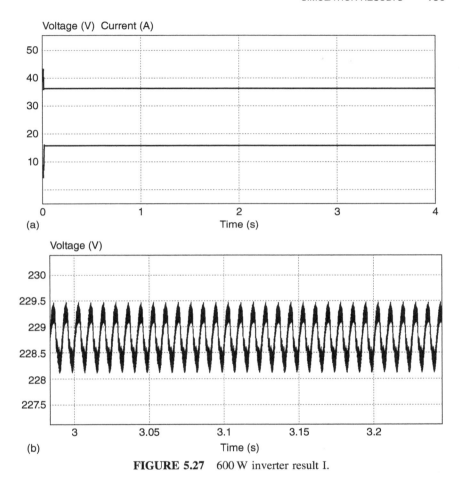

FIGURE 5.27   600 W inverter result I.

panel is supplying the power to the inverter, which can be seen by the red voltage and blue current input to the inverter in Figure 5.29a. Figure 5.29b shows the voltage output of the first stage with oscillations occurring around +/− 0.5 V.

Figure 5.30 shows the output results of the dual-mode inverter operating under normal conditions. The sensors were placed on the outputs of the filters as shown in Figure 5.23. As is expected, the inverter is outputting current through the micro-inverter portion, while the 600 W portion is open circuited.

5.4.3.2 Operation under Fault (600 W Mode)   Figure 5.31a shows the steady-state operation of the microinverter operating under 600 W mode. Both solar panels are supplying the power to the inverter, which can be seen by the red voltage and blue current input to the inverter in Figure 5.31a. Figure 5.31b shows

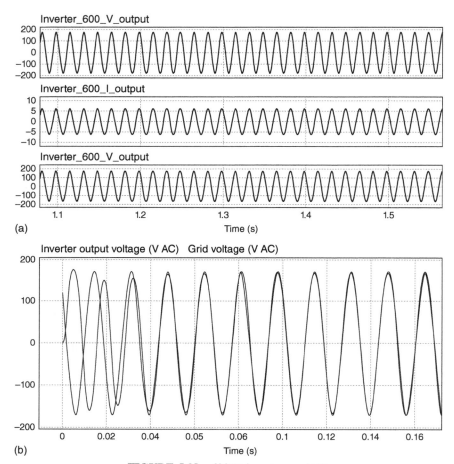

FIGURE 5.28 600 W inverter result II.

the DC link voltage after the first stage with the second-order harmonics oscillating around +/−1.2 V.

Figure 5.32 shows the inverter output results of the dual-mode inverter operating under fault conditions. As with Figure 5.30, the sensors were placed on the outputs of the filters for the readings. The top two graphs show the output of the 300 W filter, which is now open circuited. The middle graph depicts the voltage output from the 600 W filter, with output current directly underneath. Finally, the bottom-most graph shows the grid reference voltage. From the figure, it can be seen that the inverter is operating as expected, under the 600 W mode conditions.

5.4.3.3 Transient between Modes

Figure 5.33 shows the operational results of a transition between modes. At 2 s simulation time, a signal is sent

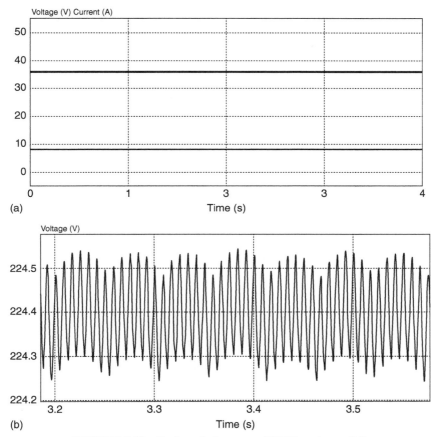

FIGURE 5.29   Dual-mode inverter – 300 W mode result I.

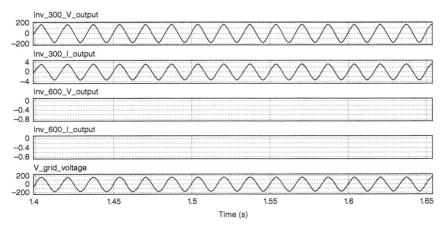

FIGURE 5.30   Dual-mode inverter – 300 W mode result II.

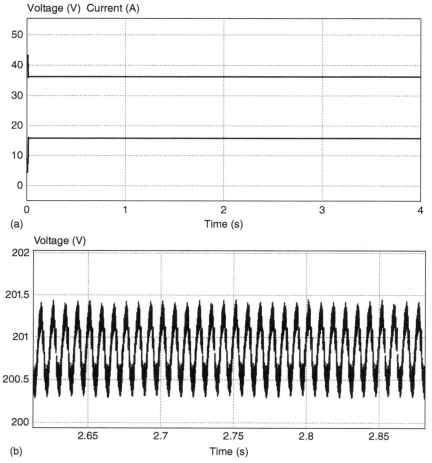

FIGURE 5.31   Dual-mode inverter – 600 W mode result I.

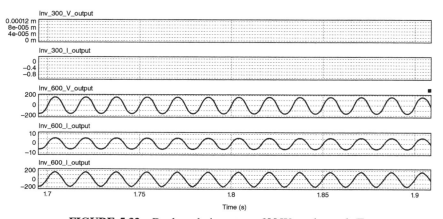

FIGURE 5.32   Dual-mode inverter – 600 W mode result II.

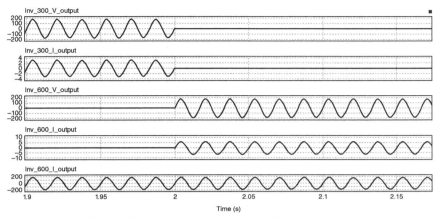

FIGURE 5.33   Dual-mode inverter: transition between modes.

connecting the second solar panel to the system. The inverter's control system recognizes the need to adjust (the need will be caused by a faulted inverter and shown in Section 5.8) and does so accordingly. The 300 W relays are opened, disconnecting those components from the circuit while closing the relays for the 600 W inverter. No transient or resynchronization is needed as PLL has already been synchronized. This allows for quick recovery and reconnection, which occurs approximately in 0.0541 s of the fault, well below the IEEE Standard of clearing within 160 ms.

5.4.4   KPI Analysis

The final section in this outlines the KPI analysis done for the inverters.

Table 5.4 summarizes the KPI results for each of the inverters designed. The microinverter exhibited the highest efficiency due to the internal resistances being lower. The dual-mode inverter operating in 300 W mode dropped the efficiency down from 95.5 to 90%; this was to be expected with the addition of the relays throughout the circuit adding to the internal resistances with in the subcircuits.

5.4.4.1   *Comparison to Equivalent Microinverters*   The following microinverter designed here has a peak efficiency of 95.5% and a THD of 2.11%. Table 5.5 summarizes existing 300 W microinverters currently on the market with their specifications from their data sheet.

The table describes the microinverter with power ranges around 300 W currently available on the market. The summary of the table shows that the preliminary microinverter design proposed by this research work has and efficiency on par with the microinverters currently on the market, but surpasses them in a lower THD. Further experimental design (hardware design) will be needed to

TABLE 5.5 Comparison of 300 W Microinverters

Microinverter (300 W)	Power (W)	MPPT (VDC)	Efficiency (%)	THD (%)	Operating Temperature (°C)
UOIT	300	25–50	95.5	2.11	N/A
Enecsys UNIV-300GTS-M	300	22–50	95.2	2.4	−40 to +85
Enphase M250	300	22–48	96.5	<5	−40 to +85
Solarbridge Pantheon	250	18–36	95.5	<5	−40 to +65
Siemens SMIINV215R60XX	270	22–36	96	3	−40 to +65

ensure calculations and simulations data are correct; however, Table 5.5 is meant to provide preliminary comparison between the designed 300 W microinverter and those available on the market.

As with the 600 W inverter, a preliminary comparison of the designed inverter is done with the ones currently available on the market. To restate, the 600 W inverter designed within this chapter has a peak efficiency of 88.5% and a THD of 1.74% (see Table 5.6).

Comparable to marketplace inverters, the preliminary 600 W inverter designed herein provides one of the highest efficiencies in its power level and surpasses with a lower THD output. N/A in Table 5.2 refers to data not given within data sheets online. As with the 300 W microinverter, further hardware design will be needed to validate both the calculations and simulations data. This table is meant to provide preliminary comparison between the designed 600 W inverter and ones currently available on the market.

5.5 MICROINVERTER SYSTEM EVALUATION

5.5.1 Key Performance Indicators

KPIs were first introduced in Section 5.3 as a means to evaluate the inverter performances. Restating, KPI is a useful performance measurement tool used to

TABLE 5.6 Comparison of 600 W Microinverters

Inverter (600 W)	Power (W)	MPPT (VDC)	Efficiency (%)	THD (%)	Operating Temperature (°C)
UOIT	600	25–50	88.5	1.74	N/A
Chaomin cm-tie 600 W	600	10.5–28	85	<5	−25 to +65 °C
Eco-Worthy GI600-24120IP65-1	600	22–50	N/A	<5	−40 to +65
i-mesh-bean SUN-600G	540	22–60	92	<5	−10 to +45
Power Jack PSWGT-600-14-28-110	600	14–28	87	2	N/A

assess the characteristics of a device, system, component, organization, and so on. For example, different control methods (fuzzy logic, PI, PID) for motor control can use KPI to assess the performance characteristics such as overshoot, response time, settling time, and efficiency of the controller. KPIs allow a common platform to compare and contrast the performance characteristics against one another. Some of the most important KPIs used to evaluate the microinverter system are shown in the following sections.

5.5.1.1 *Initial Cost System*   The initial cost of the system is the first KPI to be explored.

$$\text{Initial}_{\text{cost}}(\$) = \sum \text{Components} \tag{5.32}$$

In order to deploy the system, there are components such as the solar panels, roof mounts, microinverters, and additional components. The additional components refer to the extra components needed to operate the resiliency case study in question. A list of the extra components will be provided in each case study section.

5.5.1.2 *Yearly Revenue*   Yearly revenue is a KPI used to estimate the yearly income generated by the system.

$$R_{\text{yearly}} = \sum_{i=1}^{12} PV_{\text{potential}} \cdot P_{\text{out}} \cdot C_{\text{microfit}} \tag{5.33}$$

Here $PV_{\text{potential}}$ (kWh/kW) is the monthly PV potential data taken from the Federal Government of Canada Web site (location will be shown in Section 5.5.2), P_{out} is the power output of the entire system (kW), and C_{microfit} ($/kWh) is the government subsidy rate for roof-mounted solar panels of 29.4 cent/kWh [35].

5.5.1.3 *Payback Period (PP)*   Payback period refers to the amount of time it will take to recoup the funds from the capital investment.

$$PP = \frac{\text{Initial Cost}}{\text{Annual Cash inflow}} \tag{5.34}$$

where initial cost is the capital investment for the system and annual cash inflow is the yearly revenue generated by the system.

5.5.1.4 *Efficiency*

$$\eta = \frac{P_{\text{in}} - \sum P_{\text{losses}}}{P_{\text{in}}} \times 100 \tag{5.35}$$

Here P_{in} is the power output from the solar panel, $\sum P_{\text{losses}}$ are the losses associated with the boost converter, DC links, inverter stage, filter, and so on.

5.5.1.5 *Response Time* The following KPI is used to describe how fast the resiliency scenario is deployed in the event of an inverter failing within the system.

$$T_{\text{respon}} \ (\text{ms}) = T_{\text{res deploy}} - T_{\text{failure detected}} \tag{5.36}$$

The response time (ms) is a measure of the length of time it takes from the resiliency scenario to be deployed. Therefore, it is the time at which the resiliency scenario was deployed subtracted from the time at which failure occurred.

The KPI are the bases used to assess the economic and performance characteristics of the different case studies. A problem arises when trying to summarize the KPI in order to evaluate the system as a whole (rather than by a specific characteristic). This is because KPIs have different units depending on the characteristic in question. For example, initial cost in dollars, payback period in years, resiliency time in milliseconds, and so on. While this is a great method to compare and contrast individual characteristics, it provides a problem when looking to summarize all the characteristics for evaluation. For example, a designer might wish for emphasis on more than one specific performance criteria when designing a system. In order to overcome this problem, a novel method was designed based on turning the KPI into unitless values.

5.5.2 Per Unit Key Performance Indication

In order to provide a common platform to assess the entire system as a whole and summarize it, the KPI evaluations were had to be unitless.

5.5.2.1 *PU–KPI Governing Equation* The following provides the governing equation for the proposed method:

$$\text{KPI}_{\text{resiliency}} = \sum_{j=1}^{n} a_j \cdot \text{PU KPI}_j \tag{5.37}$$

where j is the number of KPI, n is the total number of KPI, a_j is the weighting factor assigned, and PU KPI$_j$ is the specific value from the per unit KPI calculation. The essence of the PU-KPI method is that the larger the outcome summation, the better the system.

The PU-KPI for each case are summarized into a table. These values compare the resiliency case against a system without any resiliency deployed. In addition, some of the PU-KPI show the resiliency performance changes between modes of operation, that is, the deviation of performance from operating normally to operating under fault. Each PU-KPI is multiplied by its associated weighting factor, which is determined based on the stakeholder's requirements. The more emphasis a stakeholder requires on a specific(s) performance criteria, the higher the weight attached to the PU-KPI. Upon completing the PU-KPI table, each PU-KPI for the specific case is then summed. The case that has received the highest final score shows the best method to achieve resiliency for the stakeholder based on their overall system requirements.

5.5.2.2 Per Unit Resiliency Response Time

IEEE Std. 1547 states that a disconnection of a distributed source must occur within 0.16 s (160 ms) of failure occurring [23]. Therefore, it is paramount to measure the time it takes the system to disconnect a failure and respond in kind.

$$
PU\_KPI_{\text{response time}} = \frac{\text{IEEE fault clearing standard}}{T_{\text{res deploy}} - T_{\text{failure detected}}} = \frac{160 \text{ ms}}{T_{\text{res deploy}} - T_{\text{failure detected}}}
$$

$$(5.38)$$

Using the IEEE Standard, the KPI can be normalized into a ratio that effectively describes how fast/slow the resiliency scenario detects failure, disconnects failed inverter from grid, redirects power flow, and resynchronizes to the grid. Essentially, this PU-KPI shows a ratio of resiliency response time to the IEEE Std. 1547. The lower the denominator, the quicker the response time and, consequently, the higher the PU-KPI value.

5.5.2.3 Per Unit Resiliency Power Coverage

The next measurement unit for the resiliency of the system is a ratio of how much capacity the total system has versus the capacity of the resiliency scenario to output power.

$$
PU\_KPI_{\text{Res Power coverage}} = \frac{\text{Power Output (W)}_{\text{system under fault}}}{\text{Power Output (W)}_{\text{system normal operation}}} \qquad (5.39)
$$

To illustrate this PU-KPI measurement, say there is a microinverter system power output capacity of 1.2 kW. One resiliency case study uses a single redundancy microinverter of 300 W for the entire system. Now given the event that the entire 1.2 kW system goes off-line, the resiliency capacity is, therefore, 300 W/1200 W or 1/4. Recall that for the final evaluation, the larger the final result from the scenario, the better. Therefore, a system with 1.2 kW resiliency capability would have the final score for this specific PU-KPI of 1.2 kW/1.2 kW or 1. Therefore, it would be more effective.

5.5.2.4 Per Unit Efficiency Normal Operation to Resiliency Operation

The next measurement unit is a measure of how the efficiency deviates from normal operation to when the system is operating under fault.

$$
PU\_KPI_{\text{E N to F}} = \frac{\text{Efficiency}_{\text{system under fault}}}{\text{Efficiency}_{\text{system normal operation}}} \qquad (5.40)
$$

This is used in order to help understand how the efficiency changes between operating modes of the system. In certain applications, achieving a maximum amount of efficiency may be an important factor; therefore, this PU-KPI was developed in case such a need arises.

5.5.2.5 *Per Unit Efficiency Case to Case* This PU-KPI refers to how the efficiency deviates from case 0 to the resiliency case study being evaluated.

$$PU\_KPI_{E \text{ Case to Case}} = \frac{\text{Efficiency}_{\text{system Case j}}}{Efficiency_{\text{system Case 0}}} \tag{5.41}$$

5.5.2.6 *Per Unit THD Normal Operation to Resiliency Operation* The next measurement unit is needed to measure the THD of the system operating under normal operation and its deviation to operating in the resiliency mode (under fault). Extra relays and Mosfets are included in overall circuitry to redirect power flow. These switches then introduce new resistances into the system. This can have an effect on the output filter and, therefore, have an effect on the THD of the system. Therefore, this is needed to check the deviation from normal operation to resiliency operational mode.

$$PU\_KPI_{THD} = \frac{THD_{\text{normal operation}}}{THD_{\text{under fault}}} \tag{5.42}$$

5.5.2.7 *Per Unit THD Case to Case* Similar to Section 5.2.5, this PU-KPI is how the THD deviates from case 0 to case *j*.

$$PU\_KPI_{THD \text{ Case to Case}} = \frac{THD_{\text{case 0}}}{THD_{\text{case } j}} \tag{5.43}$$

5.5.2.8 *Per Unit Initial Cost* Economic evaluations are also valuable for evaluating the system. As such, the PU-KPI was developed to show the deviation from the original initial cost of the system to the case study deployed.

$$PU\_KPI_{\text{Initial Cost}} = \frac{\text{Initial Cost}_{\text{case 0}}}{\text{Initial Cost}_{\text{case } J}} \tag{5.44}$$

The capital cost of case 0 will be less than any of the case studies. This is because all of the case studies are building upon case 0 and adding extra components. With this in mind, the initial cost of case 0 is divided by the initial cost of the case study. The greater the initial cost of case *j*, the smaller the PU-KPI becomes. Reversely, the closer the resiliency case cost is to case 0, the higher the final PU-KPI value.

5.5.2.9 *Per Unit PP* The final PU-KPI developed is related to the payback period of the system. Similar to the other PU-KPI, this shows the deviation of the

PP from the initial case to resiliency case.

$$PU\_KPI_{PP} = \frac{PP_{case\ 0}}{PP_{case\ j}} \tag{5.45}$$

Each case offers different orientation of switches and thus new power loses. Therefore, each case will have different efficiencies and consequently different yearly incomes. In addition, the initial cost differs for each case study. Thus, seeing how the PP deviates from the initial case, it is useful for evaluating the system. The formula is structured in such a way that the higher the PP of the resiliency case, the smaller and resulting PU-KPI will be. Reversely, the closer to case 0 the PP of case j, the large the resultant calculation.

Creating unitless KPI and assigning weights to each one performance criteria allows the specific requirements of the stakeholder to be emphasized. It also allows for the summation of many different KPIs to describe the overall effectiveness of different cases. For example, one designer might require the system to have a quick PP, low initial cost, and small backup capacity. Another designer might wish to have higher performance, high backup capacity, and quick response time. By tailoring the weights to meet the needs of the customer/designer, the proposed method can be customized to evaluate a wide variety of applications.

5.5.3 Resiliency Evaluation Methodology

Figure 5.34 provides the framework used for the PU-KPI evaluation method proposed.

Figure shows the framework for using the PU-KPI. The first step begins with generating case 0. The system is designed, simulated, and evaluated to provide the benchmark for the evaluation process. Following the initial design, stakeholder requirements are obtained and weights assigned to each of the requirements based on the level of need. For the sake of this work, residential PV owners and those wishing to deploy PV microinverter systems were asked to rank the PU-KPI that were of most concern to them with a number between 1 and 5, with 5 being very important and 1 having minimal importance. Based on their feedback, these were the stakeholder requirements created.

After the weights are chosen, the resiliency cases were created individually. Simulations in PSIM were carried out and the KPI outcomes were recorded down for each case during the process. It should be noted that the KPI also provides a means of feedback mechanism, ensuring that the system designed meet the IEEE grid standards. Upon completion of the different case studies, the results for all are then summarized for further use.

The next step in the framework refers to using the KPI collected and putting them into table in order to calculate the PU-KPI. Using Equations 5.7–5.14, the case study results are standardized and made unitless values. Then the values are multiplied with the specific weights chosen. With the final values calculated,

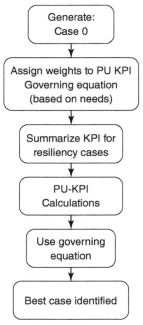

FIGURE 5.34   PU-KPI implementation framework.

the results for each specific case study are then summed up. The case with the highest final value is the case that best meets the stakeholder requirements. This process will be explained and showed further in the chapter.

5.6   CASE 0: MICROINVERTER SYSTEM

Following along with the framework described in the previous section, case 0 is generated in this section and explained.

Using the 300 W solar panels described in Section 5.2, the system consists of four 300 W Ecilpsall NRG72M solar panels (1.2 kW) arranged in the portal method as shown in Figure 5.35.

First, 1.2 kW system was chosen as the system design for a few key reasons. First, the average installation for residential solar PV in United States is 5 kW [38]. In Canada, data are more difficult to find; however, the average PV installation in Saskatchewan is 3.5 kW [39]. Therefore, 1.2 kW was chosen because it can easily be scaled to the average installation sizes. In addition to scaling up, scaling down is also a factor of importance. The power rating of 1.2 kW was chosen as the 300 W solar panels can also scale easily up to the 1.2 kW system. Therefore, 1.2 kW provides an excellent means to both scale up and scale down as needed. It will be shown later that scaling down is important for simulations and analysis. The

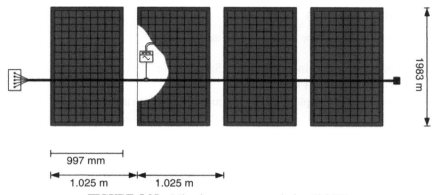

1983 m

997 mm

1.025 m | 1.025 m

FIGURE 5.35 Microinverter system design [36,37].

second key reason for choosing 1.2 kW is that one of the hypothesized methods for increasing the resiliency was to feed two solar panels to one inverter 600 W upon failure. Therefore, using a 1.2 kW size provides a good ratio 2 to 1 to ensure that the system is adequately supported in case of failure. Increasing the ratio further 3 to 1, and then to 4 to 1 would have significantly higher costs of the resiliency inverter.

The sizing of the system now complete, mounting the system was the next issue to be tackled. The method for mounting the solar panels can be seen in Figure 5.35. This is known as portal method and was chosen to connect the overall 1.2 kW system. This method was chosen as it is the most common method for connecting the system and also the recommended method as seen in the installation manuals for microinverters systems from different manufactures [36,37]. This method is preferred over the landscape method due to the fact that in the landscape method, the solar panels are rotated 90° (when compared to Figure 5.35) and the main mounting pieces run length ways through the solar panel (1.938 m long each), compared to the portal methods (997 mm) as shown in Figure 5.35. Mounting in the landscape method thus requires longer braces and more material (aluminum) and a higher cost in bracing. Finally, this method allows for room for expansion with the addition of more solar panels and systems [40] and the ability to deploy the hypothesized resiliency strategies onto it.

5.7 RESILIENCY CONTROLLER DESIGN

In order for the resiliency strategies to work, a controller is needed to govern each system design. The main tasks for a resiliency controller are: first identify a fault, disengage inverter(s), redirect power flow, and then finally to reengage the resiliency scenario to the grid. With these functionality in mind, a controller was design that can be easily deployed to all the different resiliency cases and also be easily incorporated into the onboard microcontrollers.

5.7.1 Requirements

The requirements for the controller are as follows:

- Must operate within 0.16 s (IEEE Standard, including time for switched to engage/disengage [41])
- Must be easily deployed to microcontrollers (and case designs)
- Must not remove too much efficiency form overall design (including switches losses)
- Must have minimal impact with THD of system and still meet grid standards

5.7.2 Circuit Design

In order to accomplish these requirements, the following control circuit was designed in PSIM with the ability to be easily deployed into microcontrollers.

Figure 5.36 shows the resiliency control logic designed in PSIM. The control system was designed using the C-block function within the PSIM software. This block allows for C-code to be programmed within it, and thus allows it to easily be deployed to microcontrollers and software alike. The input to the controller is the output voltage of others inverter(s). The time delays are set to 0.01 s each and act as memory storage for the controller. They provide the C-blocks with previous data points of the signal. The operating principle of the controller is such that, if all the inputs are zero (the array is zero), the output to the C-block (depicted on the left-hand side) becomes 1 and triggers opening/closure of the switches in the circuit. For all other cases, the output is 0. The output of the control circuit is then sent to the Mosfets and relays located throughout the cases (each with its own time delays). Utilizing this method, the redirection of power flow has been achieved. In addition, the device is C-code and can be easily deployed in microcontrollers. Finally, it is easily adapted to the specific design cases to be evaluated in PSIM. The effectiveness of the controller will be explored further on.

5.7.2.1 Fault   In order for the resiliency controller to be tested and be evaluated, a fault is needed. This needs to be included into the overall system

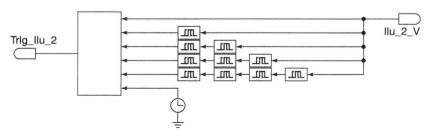

FIGURE 5.36   Control circuit for resiliency.

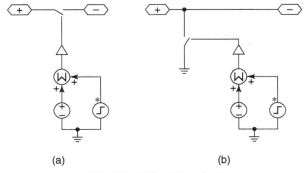

(a)   (b)

FIGURE 5.37   PSIM fault.

to test both the resiliency controller and the resiliency case study in question. Two faults were introduced: an open-circuit fault and a short-circuit fault.

Figure 5.37 shows the PSIM models for the fault to be triggered within the simulations. Figure 5.37a shows the subcircuit design used to create an open-circuit fault in the system. Figure 5.37b shows the short circuit fault created. At the top of Figure 5.37a and b, orange lines can be seen with plus and minus symbols. These indicate the connection into the overall circuit design. To place a fault, the subcircuit was put into one of the main lead wires for the microinverters and triggered by the time step of the step signal source.

5.8   RESILIENCY CASE STUDY DESIGN

The following section will first redefine the problem statement and the need for increasing resiliency to microinverter systems. Then, the assumptions will be given in order to design different case studies. Four different scenarios/cases will be shown regarding how to increase the resiliency. For each case study, the overview and operational theory will be shown. This will be followed by the PSIM simulation model created along with the operational results of the design. Finally, the case study will conclude by showing the extra components needed. The KPI for each case will be shown in the next, as it is necessary for the proposed evaluation method.

5.8.1   Need

As stated in Section 5.2, one of the key factors in the long-term success for the PV industry is the confidence of the system [42]. The reliability issues have the potential to hinder and harm the reputation of the entire industry [43]. Despite this, the industry main focus has been on reducing the overall cost and efficiency [43]. Focus is just starting to turn toward increasing the reliability, and system resiliency has often been overlooked.

In order to begin exploring the resiliency of microinverter system, is it important to first define what, in fact, resiliency is. Resiliency is defined as the ability of a system to recover from a failure [44]. From this definition, different methods were hypothesized as methods to increase the resiliency of a microinverter-based system. To keep the system operating and outputting power in the event of a microinverter failing within the system, different methods for increasing the resiliency were hypothesized. It was found that reducing energy costs of PV power configuration can be done with a common DC bus [45]. This was done for a normal PV system in order to reduce the cost. Taking this into account, possible methods to increase the resiliency in a cost-effective manner were also hypothesized.

5.8.2 Assumptions

The following section shows the assumptions needed to simulate and evaluate the microinverter systems:

- 1.2 kW system for residential microinverter applications
- 600 W maximum system will be used for simulations
- 120 V_{rms} output, portal connection
- System is for microfit in Toronto, Ontario
  - Therefore, Ontario microfit rates apply
  - Solar irradiation levels for Toronto
- Recommended spacing of solar panels as per installation manuals from Enphase

Two key points for the design and analysis that have not previously been stated are 600 W maximum system used for simulations. This is needed to cut down one simulation time and computational power. For certain case studies, simulation times can take in excess of 3 h. Increasing the software models to 1.2 kW would increase this time drastically. Therefore, 600 W output power was used for the simulation as this can provide the balance needed to scale to the 1.2 kW system and cut down on simulation times. The second point, not previously stated, is that the system is to be deployed at Toronto Ontario, and qualify for the microfit (government incentive program) in order to output its power to the electricity grid. This consequently means that solar irradiance data from Toronto, Ontario, was used (taken from government of Canada Web site [46]) along with current microfit prices for exporting roof-mounted solar panels to the grid (0.294 $/kWh [35]).

5.8.3 Case 1: Two 300 W Inverters Paired Inside Single Inverter Unit

The following section provides the first design case hypothesized in order to increase the system's resiliency.

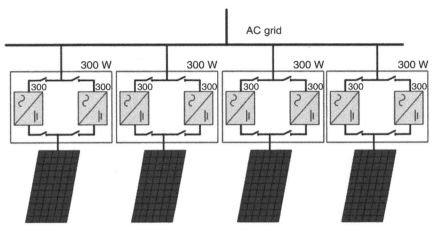

FIGURE 5.38   Case 1 overview.

5.8.3.1 *Overview Block Diagram of 1.2 kW System*   The following section provides an overview of how the full 1.2 kW system with resiliency scenario is deployed.

For the first design case proposed, each 300 W solar panel has two possible 300 W microinverters inside a single unit box mounted to one single solar pane. The first microinverter is known as the normal operation microinverter, the second being a backup or resiliency microinverter for the solar panel. This can be seen in Figure 5.38. Two 300 W microinverters are located inside the black square (representing the housing for the microinverter) and are coupled together by relays that can connect to the grid. This gives one solar panel, two possible outputs to connect to the grid.

5.8.3.2 *Operational Theory Single 300 W Inverter*   The operational theory of case 1 is described in the following section. This is done for a single microinverter with 300 W output as shown in Figure 5.39.

Figure 5.39a shows the normal operation of the microinverter. The switches consist of a combination of relays and Mosfets, which are normally closed for the main microinverter, allowing the power to flow through it. The switches for the additional backup microinverter are normally left open in order to ensure that is not coupled to the grid and panel at all. Despite the open circuit, the control system for the secondary backup microinverter is continually sensing the grid voltage. In case of fault, this allows the control system to respond and synchronize quickly. In addition to the grid, the controller is also monitoring the out-voltage for the main microinverter to know when it is necessary to engage the backup microinverter. Figure 5.39b shows a fault occurring in the normal operation of the 300 W microinverter. The output of the primary microinverter, which is being sensed by the backup microinverter, recognizes this fault occurring. The backup microinverter

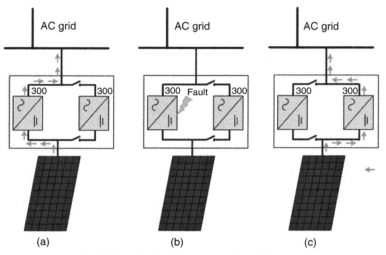

FIGURE 5.39 Case 1 operational theory.

then opens the switches to the normal operational inverter, thus removing it from the grid and solar panel. The resiliency controller within the backup microinverter then closes the switches for the backup inverter and effectively redirects the power flow through it. Thus, within the single enclosure for the solar panel, power is maintained in the event of a failure in the primary microinverter.

5.8.3.3 Circuit Design The following section outlines the PSIM model designed and simulated for case 1. In order to try and maintain a high-efficiency output, Mosfets were desired to be used to redirect power flow whenever possible. This is due to their slightly lower internal resistances (95 mΩ) [33,34] when compared to relays (100 mΩ) as well as their quicker response times [15].

Figure 5.40 shows the PSIM model for the microinverter. This is the only simulation model outputting 300 W normally. This is because each solar panel has two microinverters, which is itself a subsystem isolated from the other solar panels that contain the same resiliency strategy. The top portion of the figure shows the microinverter that is used for normal operation, the figure underneath shows the backup/resiliency microinverter. Mosfets are placed at the front end of the inverters in order to control/direct power flow into the microinverters as needed. In order to connect/disconnect the microinverter to the power grid, relays are used. Please note that in the top-most figure there is an extra blue box located near the DC link. This is the fault subcircuit and can be placed anywhere within the circuit design of microinverter 1.

5.8.3.4 Results The following section outlines the operational results of case 1 in PSIM.

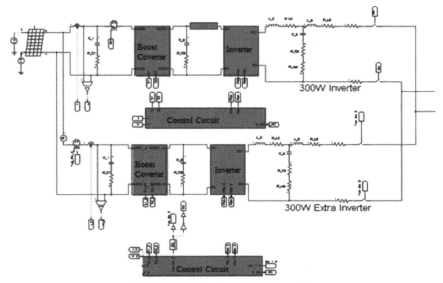

FIGURE 5.40   Case 1 PSIM model.

Normal Operation   Figure 5.41 shows the operation results for the microinverter under normal operating conditions.

The figure shows the operational results for the case 1 under normal operating conditions. The top-most graph shows the primary microinverter output voltage (i), followed by the output current directly underneath (ii). The middle graph shows the backup inverter's output voltage (iii), followed underneath by output current (iv). Finally, the bottom-most graph shows the grid reference voltage (v). It can be seen that the microinverter is synchronized with the grid, outputting its power through the primary microinverter. Microinverter 2 has no output as it is idle.

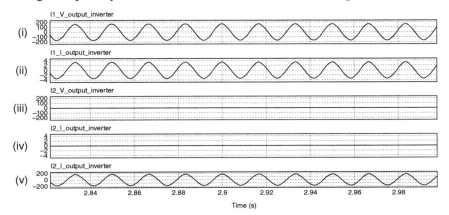

FIGURE 5.41   Case 1 results for system working under normal operation.

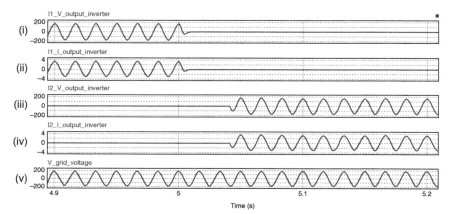

FIGURE 5.42 Case 1 microinverter 1 failure.

Inverter Failure The following sections show the output results for the system in the event of a failure in the primary operating microinverter. This is generated at 5 s simulation time.

Figure 5.42 follows the same format as Figure 5.41, with the top two graphs showing the primary microinverter's output voltage (i) and output current underneath (ii). The middle graph and the one below it show the backup microinverter's output voltage (iii) and current (iv), followed by the grid reference at the bottom (v). The fault is triggered at 5 s, which can be seen in the microinverter 1's graphs. From the figure, it can also be seen that the backup microinverter responds quickly, disconnecting the primary microinverter within a quarter cycle and overall response time is well within 160 ms.

Switching Time System output (between the two microinverters paired together) can be seen in Figure 5.43 when a fault has occurred.

From the figure, it can be seen that from the initial fault occurring, it takes approximately 0.4 s before the solar panel is outputting at full power again. This meets the initial criteria for the system to respond within 0.16 s and the system is operating as designed.

Additional Components for 1.2 kW System Scaling the system proposed by case 1 up to the full 1.2 kW, the following extra components will be needed. These extra components are additional component to case 0 in order to modify the system as shown in case 1.

A full in-depth analysis using the PU-KPI method will be conducted in Table 5.7. This table is meant to provide some closure to the case studies before the full in-depth analysis is shown. It is worth noting that for this method, the operational efficiencies have dropped from 95.5 to 94% in order to deploy the resiliency strategy. This power loss is caused by the internal resistances within the extra

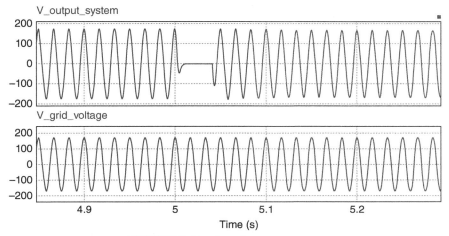

FIGURE 5.43   Case 1 system output.

switches controlling the power flow. The payback period has also increased due to the additional components. These will be shown further at the beginning of Section 5.6.

5.8.4   Case 2: Extra 300 W Microinverter in Parallel to Microinverters

The following section shows the second method hypothesized in order to increase the microinverter systems resiliency.

5.8.4.1   *Overview Circuit Block Diagram of 1.2 kW Inverter*   The following overview shows the full 1.2 kW system with case study 2 resiliency measure added to the system design.

Case 2 is similar to case 1 in the fact that it is also using an extra 300 W microinverter. The difference, however, is that in case 2, a single extra microinverter is placed in parallel to the entire 1.2 kW system rather than within every solar panel. This can be seen in Figure 5.44. The extra backup microinverter is normally disconnected from the grid. However, it is sensing the grid voltage and the outputs of all four microinverters.

TABLE 5.7   Case 1 Extra Components Invalid Source Specified

Component Description	Manufacturer	Amount Needed	Cost Per Unit ($CAN)
Relay	Songle srd-05vdc-sl-c	16	4.99
N-channel Mosfets	Sanken SKP253VR	8	2.425
300 W inverter	UOIT	4	$548.3

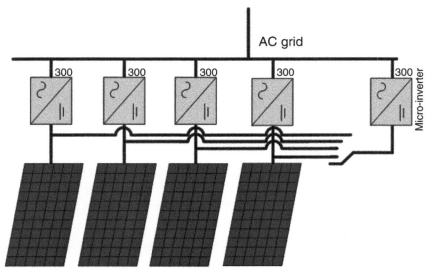

FIGURE 5.44   Case 2 overview.

5.8.4.2 Operational Theory for 600 W System   The operational theory for case 2 is shown in the following section. This is done to resemble the PSIM software in the next subsection. The system consists of two microinverters (300 W) working in parallel.

Under normal operation as shown in Figure 5.45a, each microinverter is individually connected to its respected solar panel and grid. All microinverters contain isolation switches on their inputs and outputs in order to be able to

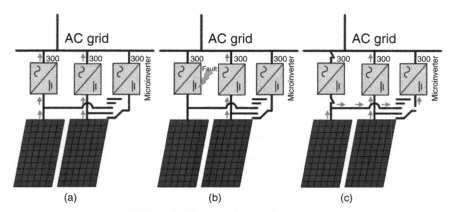

FIGURE 5.45   Case 2 operational theory.

completely isolate them if needed from both the grid and the solar. When a fault occurs, as shown in Figure 5.45b, the redundancy microinverter disconnects the failed microinverter, thereby isolating it and protecting further damage from occurring. Next, the redundancy microinverter proceeds to switch to the failed microinverter's line and connect itself to the off-line solar panel. Next, power flow is redirected through the backup microinverter. This allows the system to still have some measurement of resiliency, in the event of an inverter failure and can be seen in Figure 5.45c.

5.8.4.3 Circuit Design The following section outlines the PSIM model created for case 2. As with case 1 model, Mosfets and relays were used to control the power flow within the system.

Figure 5.46 shows the PSIM model for case 2. Two 300 W solar panels and microinverters were used to simulate the system (600 W output). Microinverter 1 and 2 are both normal operational microinverters, while microinverter 3 provides the resiliency measure for the system design. The controller for the microinverter 3 has two extra inputs in order to monitor the microinverter 1 and 2's output. And control the power flow as needed.

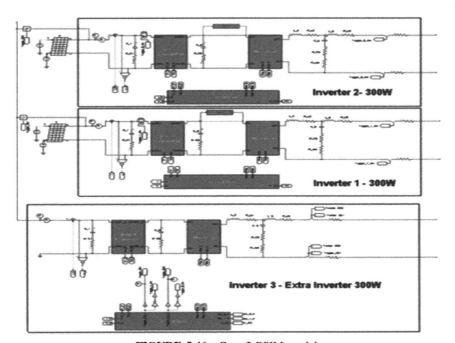

FIGURE 5.46 Case 2 PSIM model.

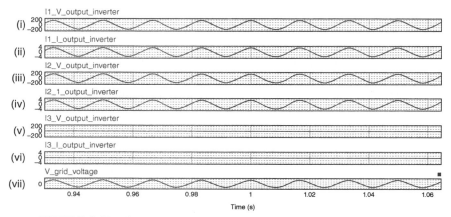

FIGURE 5.47   Case 2 results for system working under normal operation.

5.8.4.4 Results

Normal Operation   The Figure 5.47 provides the operation results for the microinverter system under normal operating conditions.

The figure demonstrates the normal operation of the microinverter system. Microinverter 1's output voltage (i) and the then current (ii) are depicted in the top two graphs. Microinverter 2's output voltage (iii) and current (iv) are depicted in graphs three and four from the top. Microinverter 3 (the backup/resiliency microinverter) output voltage (v) and current (vi) are depicted in graphs 5 and 6. Finally, the grid reference voltage is shown on the bottom-most graph (vii). Under normal operation, microinverter 3 is disconnected, which can be seen by the zero output in graphs (v) and (vi). Microinverter 1 and 2 are outputting normally and synchronized to the grid.

Inverter Failure   The following section shows the output results for the system in the event of a failure occurring in any microinverters connected to the system.

CASE 2A: INVERTER 1 FAILURE   The results shown in this section depict the system with a failure in microinverter 1. A fault is placed and triggered at 2 s simulation time.

The format of Figure 5.48 follows that of Figure 5.47, with microinverter 1's output voltage (i) and current (ii) graphs followed by microinverter 2's output voltage (iii) and current (iv). Next is the backup/resiliency microinverter's output voltage (v) and current with the grid reference voltage (vii) on the bottom-most graph. With the fault occurring at 2 s simulation time, it can be seen that microinverter 1 is disconnected within one cycle and the resiliency scenario is deployed well within 160 ms requirement.

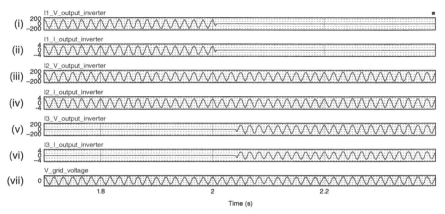

FIGURE 5.48   Case 2 microinverter 1 failure.

CASE 2B: INVERTER 2 FAILURE   The following section shows the results for a fault occurring in microinverter 2, during normal operation.

Figure 5.49 follows the same format as Figures 5.47 and 5.48. Microinverter 1's output voltage (i) and current (ii), followed underneath by microinverter's output 2 voltage (iii) and current (iv). The backup resiliency microinverter's output voltage (v) and current (vi) are then followed by the grid reference voltage (vii). The fault occurs at 2 s simulation time and the system responds as designed within roughly the same amount of time as the previous subsection. It should be noted that for this resiliency strategy, the backup microinverter can only handle 300 W. Therefore, if multiple microinverters in the system go off-line, it would not be able to handle the

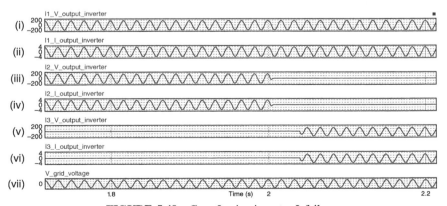

FIGURE 5.49   Case 2 microinverter 2 failure.

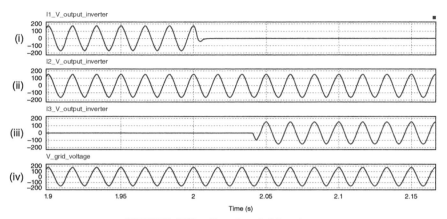

FIGURE 5.50 Case 2 switching time.

full power capability of the system. Despite this, case 2 does supply some level of resiliency to the system design.

Switching Time Figure 5.50 shows the switching time for case 2. The top-most graph depicts microinverter 1's output voltage (i), with the fault occurring at 2 s simulation time. The graph second from the top is microinverter 2's output voltage (ii). Third depicts the backup microinverter's output voltage (iii) with the grid reference voltage on the bottom-most graph (iv). From the fault occurring at 2 s simulation time, microinverter 1 is cleared, disconnected and microinverter 3 is reconnected in approximately 0.0446 s.

Additional Components 1.2 kW System Table 5.8 shows the extra components that will be needed to build case 2 resiliency strategy.

TABLE 5.8   Case 2 Extra Components Invalid Source Specified

Component Description	Manufacturer	Amount Needed	Cost Per Unit ($CAN)
300 W inverter	UOIT	1	548.3
Relay	Songle srd-05vdc-sl-c	14	4.99
Cable	Southwire RHH/RHW-2/USE-2	40 ft	0.99
Bluetooth	HC-05	5	5.48
N-channel Mosfets	Sanken SKP253VR	4	2.425

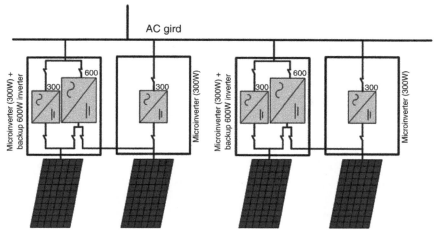

FIGURE 5.51   Overview of case 3.

5.8.5 Case 3: Backup 600 W Inverter Inside Paired Microinverters

The following section shows the third method hypothesized in order to increase the systems resiliency.

5.8.5.1 *Overview Circuit Block Diagram of 1.2 kW System*   Figure 5.51 provides the overview of the full 1.2 kW system with case study 3 resiliency measure deployed.

The figure shows the full 1.2 kW microinverter system design. For this resiliency method, solar panels (two per roof mounting rail) can be paired together and operated using a backup 600 W inverter mounted inside the microinverter housing. This provides the system with a means to output power even when a fault has occurred.

5.8.5.2 *Operational Theory for 600 W System*   The operational theory for case 3 is described in the following section. This is done for two paired solar panels that can output a maximum of 600 W. This will be shown in PSIM next.

Figure 5.52 shows the operational theory for case 3 resiliency strategy. Figure 5.52a shows the two 300 W microinverters under normal operation. Each is outputting to the grid and isolated from one another. Normally not outputting power, the 600 W inverter is monitoring the outputs of both micro-inverters as well as the grid reference voltage. Figure 5.52b shows a fault occurring in one of the microinverters. When a fault is sensed by the 600 W inverter's control system, it disconnects the switches for both microinverters. Thus, they are both taken off-line and isolated from the grid and their solar panel. Switches for the 600 W inverter are then closed, pairing the two solar panels together as the input for

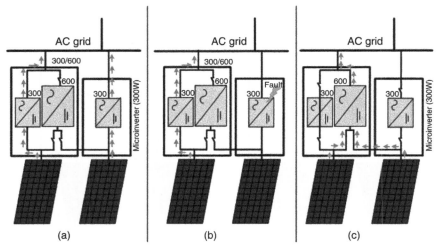

FIGURE 5.52 Case 3 operational theory.

the 600 W inverter. The 600 W is then connected to the grid, thus allowing the solar panels to continue outputting power as shown in Figure 5.52c.

5.8.5.3 Circuit Design The following section outlines the PSIM model created for case 3. As with previous cases, Mosfets and relays were applied to the circuit in order to control the power flow within the system.

In Figure 5.53, the top-most square represents is a single 300 W microinverter inside its own housing. The bottom, larger square represents the housing for the microinverter with the 600 W inverter inside.

5.8.5.4 Results The following section outlines the operational results of the case 3 created and simulated in PSIM.

Normal Operation Figure 5.54 depicts the results of the normal operation of case 3. Beginning from the top; microinverter 1's output voltage (i) and current (ii) are shown. This is followed by microinverter 2's output voltage (iii) and current (iv). Graphs 5 and 6 show the 600 W inverter's output voltage (v) and current (vi). Finally, the grid reference voltage (vii) is the bottom-most graph. Under normal operating conditions, both microinverters are directly connect to their own individual solar panel and the grid. IEEE Standards of voltage, THD, and synchronization are all met. The 600 W inverter is isolated, and waiting on standby in order to take necessary action if needed.

Inverter Failure The following section shows the results of placing a fault in the system at 1.5 s simulation time.

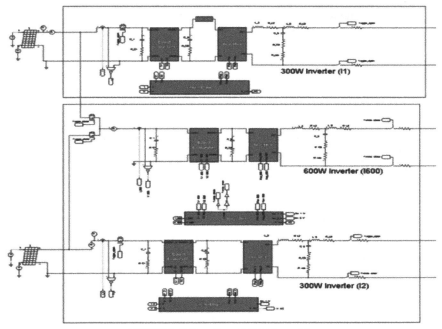

FIGURE 5.53   Case 3 PSIM model.

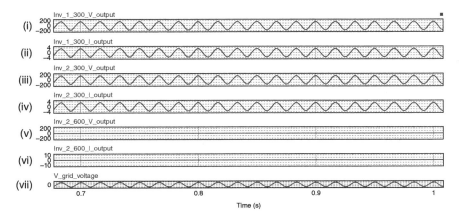

FIGURE 5.54   Case 3 normal operation.

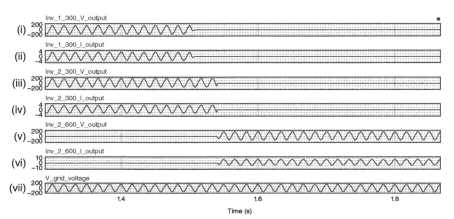

FIGURE 5.55 Case 3 microinverter 1 failure.

INVERTER 1 FAILURE   Figure 5.55 shows the results of the system (600 W) with a fault occurring at 1.5 simulation time in microinverter 1 [(i), (ii)]. A fault occurs, and the second 300 W microinverter continues to output its power [(iii) (iv)] as the fault is being recognized by the 600 W control system [(v) (vi)]. Once the fault has been recognized, the 600 W inverter opens all the switches for the microinverters and then closes the switches for the 600 W inverter. Thus, power output is maintained in the system with a small off-line period. It can be seen that the outputs are synchronized and meeting standards from the figure, and further in-depth analysis will be shown in the following section.

INVERTER 2 FAILURE   Figure 5.56 shows the results of a fault occurring in micro-inverter number 2 at 1.5 simulation time. Similar results to Figure 5.55 can be seen

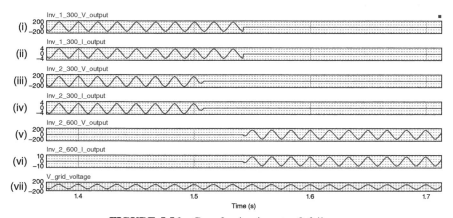

FIGURE 5.56 Case 3 microinverter 2 failure.

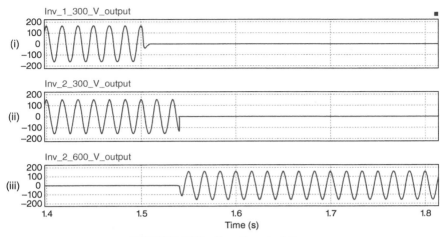

FIGURE 5.57   Case 3 control time.

and the control system is able to recognize fault in either microinverter and take the necessary corrective actions to output power.

Control Time   The following section shows the response time and output voltages of both 300 W microinverter and the 600 W inverter.

Figure 5.57 shows the response time for the resiliency system to be deployed. The fault is generated in microinverter 1 (i) at 1.5 s simulation time. From the graphs in Figure 5.57, it can be seen that once a fault has occurred in normal operation, it takes approximately 0.0467 s for the 600 W inverter (iii) to register the fault, disconnect microinverter 2, and then redirect power flow through itself.

Additional Components 1.2 kW System   Table 5.9 shows the addition components needed for case 3 to be deployed for a full 1.2 kW system.

5.8.6   Case 4: Adjustable (300–600 W) Inverters Paired

The following section provides the final case hypothesized in order to increase the systems resiliency.

TABLE 5.9   Case 3 Extra Components Invalid Source Specified

Component Description	Manufacturer	Amount Needed	Cost Per Unit ($CAN)
Extra inverter (600 W)	UOIT	2	1671.8
Relay	Songle srd-05vdc-sl-c	12	4.99
Cables for power transfer	Southwire RHH/RHW-2/USE-2	6 ft	0.99/ft
Cable for communication	Southwire RHH/RHW-2/USE-2	6 ft	0.99/ft
N-channel Mosfets	Sanken SKP253VR	8	2.425

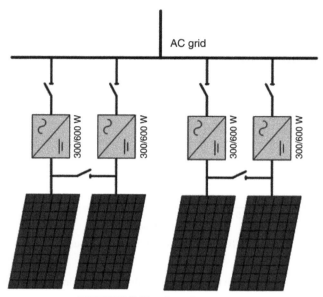

FIGURE 5.58   Case 4 overview.

5.8.6.1   *Overview Circuit Block Diagram of 1.2 kW System*   Case 4 overview can be seen in Figure 5.58. For the final case, each microinverter has been changed to the dual-mode inverter as shown in Part I of this book. Recall this inverter can modify its output to two different power ratings: 300 and 600 W. The two inverters are paired together on a single solar panel roof mount. Under normal operating conditions, each inverter is operating in the 300 W mode, connected to its own individual solar panel and grid. As such, the solar panels are isolated from one another. In the event of a failure in one of the dual-mode inverters, the working inverter adjusts itself from 300 W mode to 600 W mode. It then pairs the solar panels together as the input power supply.

5.8.6.2   *Operational Theory System 600 W*   The operational theory is described in the following section. This is done strictly for two dual-mode inverters working on the single roof mount with a maximum output power of 600 W for the system.

Figure 5.59a shows the normal operation of the system design. Under normal operation, both dual-mode inverters are operating independently in 300 W mode and exporting the solar panels power to the grid. When a fault occurs, in either of the microinverters as shown in Figure 5.59b, the faulted dual-mode inverter is disconnected from the grid. The operational dual-mode inverter, sensing that its partnered inverter is not operational, closes the switches pairing the two solar panels together. In addition, its internal components are adjusted. Both solar

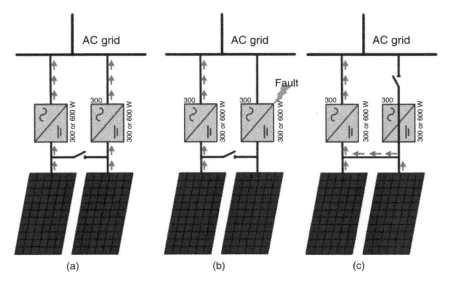

FIGURE 5.59 Case 4 operational theory.

panels are then fed into the operating dual-mode inverter as shown in Figure 5.59c.

5.8.6.3 Circuit Design Case 4 PSIM model of the dual-mode inverter is shown in Figure 5.60. As with the other cases, the limit for the simulation will be a 600 W output power maximum in order to cut down on simulation time.

The two solar panels are paired together by wires controlled by relays, and power flow into the dual-mode inverter is done with Mosfets. Inverter 1 is the top-most inverter, while Inverter 2 is the bottom-most inverter for this simulation. The results of the system will be explored in the following section.

5.8.6.4 Results The following section outlines the operational results of the final simulation created in PSIM.

Normal Operation The first section shows the simulation results for normal operation of the inverters with both dual-mode inverters operating in 300 W mode.

Figure 5.61 shows the results for the simulation run in PSIM. Dual-mode inverter 1's 300 W mode output voltage (i) and current (ii) are followed by the 600 W mode output voltage (iii) and current (iv). Dual-mode inverter 2's 300 W mode output voltage (v) and current (vi) followed by its 600 W modes output voltage (vii) and current (viii). Finally, the bottom-most graph shows the grid reference voltage (ix). From the figure, it can be seen that the system is operating as expected, both dual-mode inverters are operating in 300 W mode and synchronized

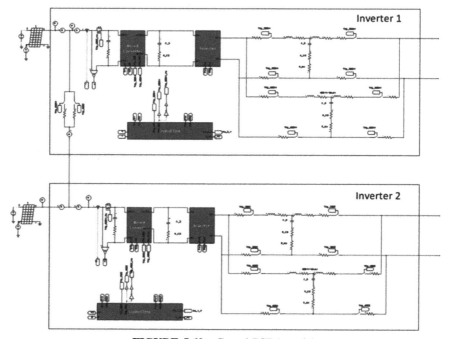

FIGURE 5.60 Case 4 PSIM model.

to the grid with low THD. It should be noted that under normal operation, the microinverter that initially has high efficiency of 95.5% has now been reduced down to 90% by the addition of the Mosfets and switches added into the overall circuit design. The internal resistances of these devices resulted in higher power loses for the circuit design.

Inverter Failure The following section provides an overview of the results when a fault is placed into one of the inverters at 2 s simulation time.

INVERTER 1 FAILURE: INVERTER 2 TURNS INTO 600 W The resiliency testing begins with placing a fault in Inverter 1, the system results of which can be seen in Figure 5.62.

A fault was placed at 2 s simulation time. In order to save space, only voltages were displayed for the results section in case 4. From top to bottom, the graphs are: dual-mode inverter 1 300 W mode (i) followed by its 600 W mode (ii); dual-mode inverter 2 300 W mode output voltage (iii) followed by its 600 W output voltage (iv); finally, grid reference voltage (v) is the bottom-most graph. The fault was placed within inverter 1 on the DC link side of the inverter stage. Dual-mode inverter 2 continues to operate in the 300 W mode. Once its control system

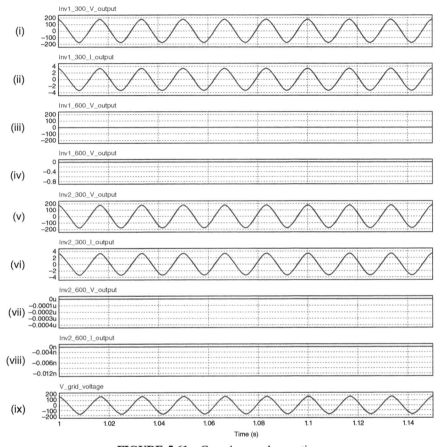

FIGURE 5.61   Case 4 normal operation.

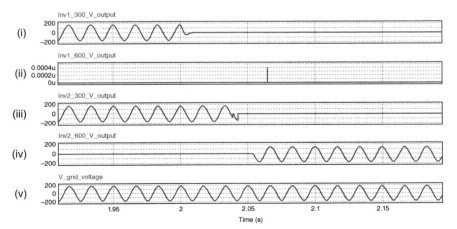

FIGURE 5.62   Case 4 inverter 1 failure.

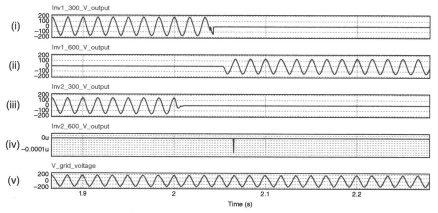

FIGURE 5.63 Case 4 inverter 2 failure.

recognizes that a fault has occurred, the necessary switches are opened isolating dual-mode inverter 1. The internal components for dual-mode inverter 2 are then adjusted in order to output 600 W. Dual-mode inverter 2 then pairs both solar panels and reconnects to the grid to resume outputting power. It can be seen that the dual-mode inverter operates effectively and within the 160 ms target.

INVERTER 2 FAILURE: INVERTER 1 TURNS INTO 600 W   The result of testing for a fault in dual-mode inverter 2 can be seen in Figure 5.63.

Similar to the previous simulation, the fault was placed a 2 s simulation time. Figure 5.63 follows the same format as Figure 5.62, with dual-mode inverter 1 300 W mode voltage (i) followed by its 600 W mode (ii). Dual-mode inverters 300 W output (iii) is then followed by its 600 W output (iv) with the grid reference occurring in the bottom-most figure (v). Other modes and current were omitted in order to try and provide a clear view of the functioning output. The fault occurs in dual-mode inverter 2 at 2 s and it is disconnected from the grid. Dual-mode inverter 1 continues to output as its control system is recognizing the failure in dual-mode inverter 2. Once the control system has recognized the fault, it modifies the switches throughout the circuit accordingly and power output continues.

Control Time   The Figure 5.64 shows the system output (the output from both inverters) for the inverters in case of a fault occurring in the second dual-mode inverter.

In the figure, (i) is dual-mode inverter 1 300 W voltage, (ii) dual-mode inverter 1 600 W voltage, and (iii) dual-mode inverter 2 300 W output voltage. The fault occurs in the secondary dual-mode inverter as the system operates under normal conditions. From Figure 5.64, it can be seen that the system takes approximately 0.0541 s from the time initial fault occurs before the system is outputting at full power again.

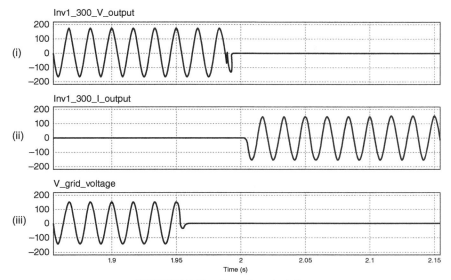

FIGURE 5.64   Case 4 control time.

Extra Components 1.2 kW System   Scaling the system to the full 1.2 kW, Table 5.10 shows the extra components that will be needed to construct the system with the proposed resiliency strategy.

5.9   RESULTS

5.9.1   Summary of KPU

Upon completion of the case studies, Table 5.11 summarizes the KPI for each case study.

Table 5.11 shows the first step as part of the PU-KPI method proposed. It is also beneficial as it summarizes the performance characteristics for each case study proposed. Cases 1–3 exhibited the same efficiency as the same configuration of

TABLE 5.10   Case 4 Extra Components Invalid Source Specified

Component Description	Manufacturer	Amount Needed	Cost Per Unit ($CAN)
Relay	Songle srd-05vdc-sl-c	4	4.99
Cables (power)	Southwire RHH/RHW-2/USE-2	10 ft	0.99/ft
Cables (communication)	Southwire RHH/RHW-2/USE-2	10 ft	0.99/ft
Mosfets	Sanken SKP253VR	4	2.54

TABLE 5.11 KPI Case Study Table

Key Performance Indicator	Case 0: No Resiliency Added	Case 1: Two 300 W Inverter Inside	Case 2: 300 W Inverter in Parallel	Case 3: 600 W Inverter Inside	Case 4: Dual-Mode Inverter
Peak operational efficiency	$P_{in}=287.44$ W $P_{out}=274.44$ W $\eta=95.47\%$	$P_{in}=287.44$ W $P_{out}=270.16$ W $\eta=94\%$	$P_{in}=287.44$ W $P_{out}=270.16$ W $\eta=94\%$	$P_{in}=287.43$ W $P_{out}=269.5$ W $\eta=94\%$	$P_{in}=255.3$ W $P_{out}\,287.43$ $\eta=88.8\%$
Peak operating THD	2.11E-002	2.04518E-002	2.04518E-002	2.03612E-002	2.4E-002
Overall Resiliency Response time	N/A	0.03945 s	0.0446 s	0.0467 s	0.05417 s
Peak operational efficiency under fault	N/A	$P_{in}=287.44$ W $P_{out}=270.16$ W $\eta=94\%$	$P_{in}=276.21$ W $P_{out}=256.59$ W $\eta=92.9\%$	$P_{in}=565.4$ W $P_{out}=482.82$ W $\eta=85.4\%$	$P_{in}=559.7$ W $P_{out}=415.3$ W $\eta=74.2\%$
Operating THD under fault	N/A	2.04518E-002	1.94707E-002	1.7929E-002	1.3556E-002
Overall Cost for 1.2kW system	$3871.64	$6168.56	$4560.62	$7320.88	$10,172.34
Estimated yearly energy generated	1315.2 kWh	1296 kWh	1296 kWh	1296 kWh	1224 kWh
Estimated Yearly Income	$387.98	$382.32	$382.32	$382.32	$361.1
Payback period for 1.2kW system	9.98 years	16.13 years	11.92 years	19.12 years	28.2 years

TABLE 5.12   PU-KPI Results Table

PU-KPI	a_j Weight	Case 1	Case 2	Case 3	Case 4
PU_KPI$_{\text{Resnse Time}}$	1	4.06	3.59	3.43	2.95
PU_KPI$_{\text{ResPower overage}}$	1	1.00	0.25	1.00	1.00
PU_KPI$_{\text{E N to F}}$	1	1.00	0.99	0.91	0.84
PU_KPI$_{\text{E Case to Case}}$	1	0.98	0.98	0.98	0.93
PU_KPI$_{\text{THD}}$	1	1.00	1.05	1.14	1.76
PU_KPI$_{\text{THD Case to Case}}$	1	1.03	1.03	1.04	0.88
PU_KPI$_{\text{Initial Cost}}$	3	1.88	2.55	1.59	1.14
PU_KPI$_{\text{PP}}$	4	2.47	3.35	2.09	1.42
$\sum_{j=1}^{n} a_j \cdot \text{PU KPI}_j$		13.43	**13.79**	12.17	10.91

switches was used in order to control a redirect power flow. The THD remain low ranging from 2 to 2.4% for all case studies. All case studies exhibited lower efficiency while operating under faulted conditions as an addition switch or two was needed. The exception to this is case 1, which has the same efficiency during both modes of operation.

5.9.2   Calculating and Mapping of PU-KPI

Applying Table KPI values through the proposed PU-KPI formulas as shown in Equations 5.7–5.14, Table 5.12 can be created from the results.

Table 5.12 shows the results of the PU-KPI for each case as well as the weighing factor requested by the stakeholder. The stakeholder requested a higher emphasis on the system having a low initial cost and a quick payback period. Each weighing factor was multiplied by the corresponding case study PU-KPI value on the same row as it. The bottom-most column sums the multiplication of the two values.

The case that best suits the stakeholder's request is case 2: 300 W microinverter in parallel. Case 2 shows as having the best economic performance and perform- ance characteristics among the four different case studies. It should be noted that if more backup power coverage is requested, and a higher emphasis put on that, then case 1 would have been the best method in order to deploy resiliency as the two final scores are very close to one another.

5.10   CONCLUSION

This chapter presented three different inverter design cases developed and simulated using readily available components. The inverters were designed based on calculations and then modeled in PSIM. The design phase was described for the inverter's different subcircuits, which were then integrated together for the overall final design. Inverters with two different power levels were designed. The first, a 300 W microinverter was designed and exhibited a high efficiency of 95.5% with a low THD of 2.1%. Second, a 600 W inverter was designed using the same subcircuit topology. The preliminary design for the 600 W inverter was found to have an efficiency of 88.5% and a low THD of 1.74%. Finally, a third inverter was designed, called the dual-mode inverter, which can modify its internal structure in order to output at two different power levels based on the need. Section 5.8 will apply these inverters in order to increase the resiliency of microinverter systems for residential applications.

This chapter focused on improving the resiliency of microinverter systems for single-phase grid-connected applications. A novel method for evaluating microinverter systems was introduced. This method is based on KPI and can easily be tailored in order to meet the specific design requirements for the systems. A resiliency controller was designed based on C-code. Finally, four different methods for increasing the resiliency of microinverter-based systems were proposed and evaluated. Simulation for the systems were carried out in PSIM software. The addition of an extra backup microinverter in parallel to the system was found to be the most effective method to increase the resiliency of a microinverter system based on having low initial cost and low payback period.

REFERENCES

1. International Energy Agency, (2014) Technology Roadmap: Solar Photovoltaic Energy – 2014 Edition, Available at https://www.iea.org/publications/freepublications/publication/technology-roadmap-solar-photovoltaic-energy—2014-edition.html (accessed September 30, 2015).

2. National Resources Canada (2014) Improving Energy Performance in Canada – Report to Parliament Under the Energy Efficiency Act, Available at http://oee.nrcan.gc.ca/publications/statistics/parliament10-11/5.cfm?attr=0. (accessed March 12, 2015).

3. Government of Canada (2015) National Resources Canada: Photovoltaic Technology Status and Prospects: Canadian Annual Report 2014, Available at http://www.nrcan.gc.ca/energy/publications/sciences-technology/renewable/solar-photovoltaic/16384. (accessed Sepetember 30, 2015).

4. National Renewable Energy Laboratory (2006) A Review of PV Inverter Technology Cost and Performance Projections, US. Department of Energy, Battelle.

5. van Sark, W., Muizebelt, P., Cace, J., de Vries, A., and Rijk, P. (2014) Price development of photovoltaics modules, inverters, and systems in The Netherlands in 2012. *Renewable Energy*, 71 (11), 18–22.

6. Sandia National Laboratories (2011) Utility-Scale Grid Tied PV Inverter Reliability Workshop Summary Report, U.S. Department of Commerce, Springfield.

7. Mohan, N., Undeland, T., and Robbins, W. (2003) *Power Electronics, Converters, Applications, and Design*, John Wiley & Sons, Inc., New Jersey.

8. Owen, E. (1996) "History [origin of the inverter]," IEEE Industry Applications Magazine, pp. 64–66.

9. Floyd (2007) *Electronic Devices: Conventionals Current Version*, Prentice Hall, Upper Saddle River, NJ.

10. Floyd (2006) *Principles of Electric Circuit: Conventional Current Version*, Prentice Hall, Upper Saddle River, NJ.

11. Kjaer, S. and Pedersen, J.B.F. (2005) A review of single-phase grid connected inverters for photovoltaic modules. *IEEE Transactions on Industry Applicationcs*, 41 (5), 1292–1306.

12. Tomova, A., Petkova, M., Antchev, M., and Antchev, H. (2013) Computer investigation of a sine- and cosine-based phased-locked loop for single phase grid connected inverter. *International Journal of Engineering Research and Applications*, 3 (1), 726–728.

13. Nambodiri, A. and Wani, H. (2014) Unipolar and biopolar PWM inverters. *International Journal for Innovative Research in Science and Technology*, 1 (7), 237–243.

14. Bo, L., Guo-Chun, X., and Lu, W. (2013) Comparison of performance between bipolar and unipolar double-frequency sinusoidal pulse width modulation in a digitally controlled H-bridge inverter system. *Chinese Physics B*, 22 (6), 1–8.

15. Muhammad, R. (2007) *Power Electronics Handbook*, Elsevier, San Diego, CA.

16. Guo, X., Wu, W., and Gu, H. (2011) Phase lock loop and synchronization methods for grid interfaced converters: a review. *Przeglad Elektrotechniczny*, 87 (4), 182–188.

17. Kroupa, V. (2003) *Phase Lock Loops and Frequency Synthesis*, John Wiley & Sons, Inc., New York.

18. Carranza, O., Trujillo, C., Ortega, R., and Rodrigueze, J. (2013) Synchronization to the grid using linear kalman filter applied to single phase inverters. Calidad de la Energia Electrica, Medellin.

19. Brasil, T. (2014) Comparative study of single-phase PLLs and fuzzy based synrchronism algorithm, IEEE Industrial Electronics, Istanbul.

20. Thacker, T., Wang, R., Dong, D., Burgos, R.W., and Boroyevich, D. (2009) Phase-locked loops using state variable feedback for single-phase converter systems. Applied Power Electronics Conference and Exposition, Washington.

21. Guerrero-Rodriguez, N., Rey-Boue, A., Rigas, A., and Kleftakis, V. (2014) Review of synchronization algorithms used in grid-connected renewable agents. International Conference on Renewable Energies and Power Quality, Cordoba.

22. Karimi-Ghartemani, M. (2012) A unifying approach to single-phase synchronous reference frame PLLs. *IEEE Transaction on Power Electronics*, 28 (10), 4550–4556.

23. Antchev, M., Pandiev, I., Petkova, M., Stoimenov, E., Tomova, A., and Antchev, H. (2013) PLL for single phase grid connected inverters. *International Journal of Electrical Engineering and Technology*, 4 (5), 56–77.

24. IEEE (2003) Standard for Interconnecting Distributed Resources with Electric Power Systems, IEEE, Piscataway.

25. Eclipsall Energy Group (2015) Products Available at http://www.eclipsall.com/ (accessed Janruary 5, 2015).

26. Chen, R., Liu, Y., and Peng, F. (2014) DC capacitor-less inverter for single-phase power conversion with minimum voltage and current stress. *IEEE Transactions Power Electronics*, 30 (10), 5499–5507.

27. El-Hefnawi, S. (1997) Digital firing and digital control of a photovoltaic inverter. *Renewable Energy*, 12 (3), 315–320.

28. Cha, H. and Vu, T. (2010) Comparative analysis of low-pass output filter for single-phase grid-connected photovoltaic inverter. Applied Power Electronics Conference and Exposition (APEC), Palm Springs.

29. Wu, W., Huang, M., and Blaabjerg, F. (2014) Efficiency comparison between the LLCL and LCL filters based single phase grid tied inverters. *Archives of Electrical Engineering*, 63 (1), 63–79.

30. Lang, Y., Xu, D., Hadianamrei, S., and Ma, H. (2005) A novel method of LCL type utility interface for three-phase voltage source rectifier. Power Electronic Specialists Conference, Recife.

31. Dolara, A., Faranda, R., and Leva, S. (2009) Energy comparison of seven MPPT techniques for PV systems. *Journal of Electromagnetic Analysis and Applications*, 1 (3), 152–162.

32. Gomes de Brito, M., Galotto, L., Sampaio, L., Azevedo e Melo, G., and Cansin, C. (2013) Evaluation of the main MPPT techniques for photovoltaic applications. *IEEE Transactions on Industrial Electronics*, 60 (3), 1156–1167.

33. Digikey, (2015) Parts, Available at http://www.digikey.ca/ (accessed September 20, 2015).

34. Mouser Electronics, (2015) Parts. Available at http://ca.mouser.com/ (accessed September 20, 2015).

35. IESO (2015) microFit. Available at http://microfit.powerauthority.on.ca/types-renewable-technology-types. (accessed September 1, 2015).

36. Enphase Energy (2015) Installation and Operations Manual, Enphase, Petaluma.

37. Siemens (2011) Siemens micro-inverter system: selection and application guide, Oakville.

38. Solar Energy Industries Association (2015) Solar Photovoltaic Technology. Available at http://www.seia.org/research-resources/solar-photovoltaic-technology (accessed November 1, 2015).

39. White, S., (2011) Is this solar-power program a money-saver? The Globe and Mail, November 30. Available at http://www.theglobeandmail.com/globe-investor/personal-finance/household-finances/is-this-solar-power-program-a-money-saver/article4180338/ (accessed November 2, 2015).

40. Ribeiro, L., Saavedra, O., Lima, S., de Matos, J., and Bonan, G. (2012) Making isolated renewable energy systems more reliable. *Renewable Energy*, 45 (1), 221–231.

41. IEEE Std. (2003) Interconnecting Distributed Resources with Electric Power Systems, IEEE, Piscataway.

42. National Renewable Energy Laboratory (2006) A Review of PV Inverter Technology Cost and Performance Projections, US. Department of Energy, Battelle.

43. Sandia National Laboratories (2011) Utility-Scale Grid Tied PV Inverter Reliability Workshop Summary Report, U.S. Department of Commerce, Springfield.

44. Hollnagel, E. and Nemeth, C. (2009) *Resilience Engineering Perspectives, Volume 2: Preparation and Restoration*, CRC Press, Boca Raton, FL.

45. He, F., Zhao, Z., and Yuan, L. (2012) Impact of inverter configuration on energy cost of grid-connected photovoltaic systems. *Renewable Energy*, 41 (5), 328–335.

46. Government of Canada, (2015) Photovoltaic potential and solar resource maps of Canada, Natural Resources Canada. Available at http://pv.nrcan.gc.ca/ (accessed September 1, 2015).

PART III

ENERGY CONSERVATION STRATEGIES

CHAPTER 6

INTEGRATED PLANNING AND OPERATIONAL CONTROL OF RESILIENT MEG FOR OPTIMAL DERs SIZING AND ENHANCED DYNAMIC PERFORMANCE

HOSSAM A. GABBAR,[1,2] AHMED M. OTHMAN,[1,3] and ABOELSOOD ZIDAN[1]

[1]Faculty of Energy Systems and Nuclear Science, University of Ontario Institute of Technology, Oshawa, Canada
[2]Faculty of Engineering and Applied Science, University of Ontario Institute of Technology, Oshawa, Canada
[3]Electrical Power and Machines Department, Faculty of Engineering, Zagazig University, Zagazig, Egypt

6.1 INTRODUCTION

Microenergy grid (MEG) is a relatively small-scale localized energy network with energy sources and loads. MEG has two modes of operation: grid-connected mode as it is connected to the traditional centralized grid; islanded mode as it works autonomously due to physical and/or economic conditions. A MEG requires minimum energy transmission from/to remote regions as it has its own local energy sources. It is expected that future distribution system will be composed of a number of interconnected MEGs to reduce energy loss, cost of a transmission network, and the risk of energy supply failure. Due to their storage systems, MEGs have positive impact in increasing the penetration level of renewable energy sources. For enhanced reliability, MEGs depend on multiple fuel-based generators such as diesel, natural gas, and renewable sources (solar and wind). As distributed energy resources (DERs) are located close to loads in MEGs, the wasted heat from these DERs can be used in covering the local heat demand [1–3].

Energy Conservation in Residential, Commercial, and Industrial Facilities, First Edition.
Edited by Hossam A. Gabbar.
© 2018 The Institute of Electrical and Electronics Engineers, Inc. Published 2018 by John Wiley & Sons, Inc.

Natural gas is considered as the primary fuel for combined heat and power (CHP) units. CHP and renewable energy are key clean energy sources in MEGs. Thus, planning of CHP and renewable energy sources represents a vital step to realize and guarantee self-sufficient, stable, and efficient MEG performance. MEGs that integrated with CHP and renewable energy sources have many positive aspects such as increased reliability, reduced wasted energy, reduced emissions, and reduced capital and operational costs [4,5]. Due to the technical challenges such as flow control and reliability issues, MEG requires an efficient and adaptive control scheme. Numerous developed devices can be installed to manage the technical impacts. Power electronic converters and inverters with switching capability can be used to allow controllable actions and operation of AC/DC systems [6–8]. Developed flexible AC transmission systems (D-FACTS) allow fast and accurate control actions to enhance the system dynamics. Green plug-energy economizer (GP-EE) is a type of D-FACTS that can be installed with MEG for enhancing its performance. GP-EE can be applied for improving power quality aspects such as voltage regulation improvement, power factor improvement, and transient content minimization in voltage and current waves.

Modern heuristics techniques have many applications in various areas of power systems. Adaptive controllers and optimizers can be compared with conventional ones to show their effectiveness. Enhanced heuristic optimization method as backtracking search algorithm (BSA), which is a recent evolutionary technique, that searches the global optimal solutions inside the problem space can be applied. BSA realizes remarkable impacts in solving different numerical optimization problems [9,10]. BSA will be applied to control the parameter settings of GP-EE to fine-tune the system dynamic response.

In this chapter, an integrated planning and operational control of a MEG has been presented. The planning stage aims for optimal deployment of DERs and CHPs within microgrids by selecting their capacity sizes and types. The objective is to simultaneously minimize the total net present cost and CO_2 emissions. A multiobjective genetic algorithm (GA) is proposed in that stage to handle the planning problem. The constraints involve power and heat demands constraints. Candidate technologies include CHPs with various characteristics, boilers, renewable generators, and a main power grid connection. The results of the planning stage will be fed to the operational stage to ensure that it works on the optimal patterns of the DERs. The operational stage concerns with the operational control aspect. GP-EE has been applied to enhance the performance of the MEG. A novel control adaptation for error-driven dynamic multiloop regulator has been applied. The PWM-optimized pulsing sequence for the GP-EE utilizes a triloop dynamic error-driven weighted modified PID controller. BSA will adapt the dynamic gains of PID regulation triloop. BSA will be able to achieve the desired performance of the system when it is well designed and tested to emphasize the proper selection and control of GP-EE parameters.

The proposed GP-EE schemes are very effective in enhancing power quality, improving power factor, reducing feeder losses and stabilizing the voltage profile, and mitigating harmonic distortion.

6.2 MEG DESIGN WITH ESCL DEMONSTRATIONS

Figure 6.1 shows that MEG configuration included GP-EE schemes for both AC and DC terminals. MEG has different AC and DC DER units that supply different AC and DC loads. Based on the output of the planning stage, the AC sources are DFIG wind turbine generator and microgas turbine generator. The DC sources are battery, fuel cell stack based on hydrogen, and photovoltaic arrays. The design of microgrid will include boost converters (AC/DC, DC/AC, and DC/DC converters) for reliable operation.

The CHP generators will be presented with various characteristics, boilers, thermal storage, renewable generators (wind and photovoltaic), and a main power grid connection. The candidate technologies included are natural gas turbine (NGT), hydrogen fuel cell (H_2FC), natural gas fuel cell (NGFC), wind turbine (WT), and photovoltaic (PV). NGT, H_2FC, and NGFC are assumed to operate as CHPs to generate heat and power with fixed heat-to-power ratio. The load profile

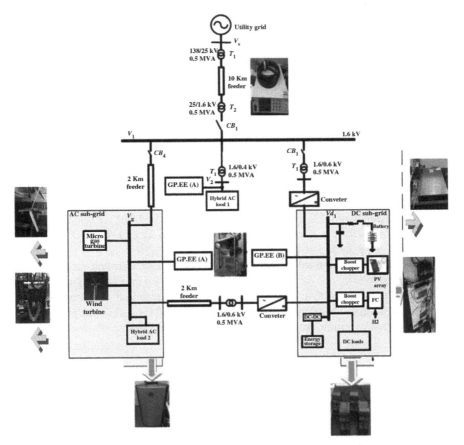

FIGURE 6.1 Configuration of MEG included GP-EE schemes.

for a year is divided into four seasons; each season was represented by 1 day with total sampling hours equal to $4 \times 24 = 96$ h/year. Various types of DC and AC loads are represented, such as resistive loads, motorized DC series motor loads, linear AC loads, nonlinear AC loads, and three-phase induction motorized loads.

The design MEG with hardware demonstrations is presented in Energy Safety and Control Laboratory (ESCL), University of Ontario Institute of Technology (UOIT). Starting from 2011, the members of ESCL are working on MEG with more than 5 projects and 20 publications. The data and analysis are simulating UOIT models and Geographical Information System, as indicated in Figure 6.2.

6.2.1 The Planning Stage

A DER model can be represented by the following equations:

$$f_{\text{fuel},i}(t) = \frac{P_i(t)}{u_i \eta_{i,\text{P}}}, \quad \forall t \in T, \quad 0 \le P_i(t) \le P_{\text{r},i}, \quad i \in G \tag{6.1}$$

$$H_i(t) = P_i(t) \frac{\eta_{i,\text{H}}}{\eta_{i,\text{P}}}, \quad \forall t \in T, \quad i \in G \tag{6.2}$$

$$E_i(t) = K_i u_i f_{\text{fuel},i}(t), \quad \forall t \in T, \quad i \in G \tag{6.3}$$

$$P_{\text{loss},i}(t) = \frac{P_i(t)}{\eta_{i,\text{P}}}(1 - \eta_{i,\text{P}} - \eta_{i,\text{H}}), \quad \forall t \in T, \quad i \in G \tag{6.4}$$

where G is the set of DERs {NGTs, NGFCs, and H_2FCs}; $f_{\text{fuel},\,i}$ is the consumed fuel by DER i at hour t; P_i is the output power from DER i at hour t; T is the set of hourly periods (i.e., 8760 for the entire year); u_i is the energy density of the fuel consumed by DER i in kWh/kg; $\eta_{i,\text{P}}$ is the power efficiency of DER i; $P_{\text{r},i}$ is the rated power for DER i; H_i is the generated heat power from DER i at hour t; $\eta_{i,\text{H}}$ is the heat efficiency of DER i; E_i is the CO_2 emissions from DER i at hour t; K_i is the carbon footprint for the energy produced by DER i in kg CO_2/kWh; $P_{\text{loss},i}$ is the power losses for DER i at hour t.

Based on seasons, the entire year is divided into four seasons and each season is being represented by one sample day, which is subdivided into 24 h segments, with total equal to $4 \times 24 = 96$ h. The total cost (capital, operational, and maintenance) and CO_2 emission are calculated. It will change based on daily load profile. The multiobjective planning problem can be defined as follows:

a. *Objective (Fitness) Function:*

$$\text{Minimize}(\text{OF}_1, \text{OF}_2) \tag{6.5}$$

$$\text{OF}_1 = f_{\text{cap}} + \sum_{t \in T}(C_{\text{ope}}(t) + C_{\text{pur}}(t)) \tag{6.6}$$

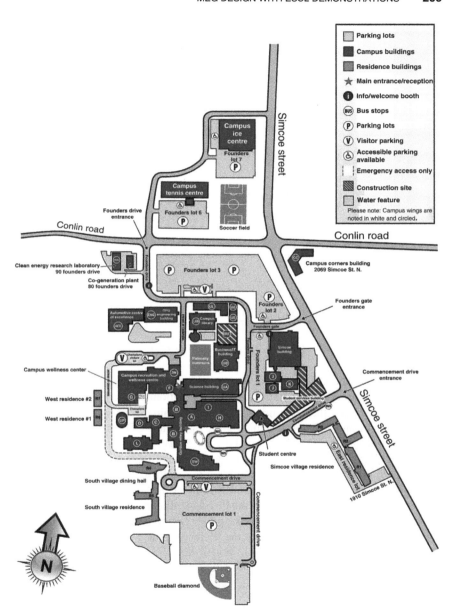

FIGURE 6.2   UOIT geographical information system.

$$OF_2 = \sum_{t \in T} (E_{grid}(t) + E_{DER}(t) + E_{bo}(t)) \tag{6.7}$$

$$f_{cap} = \sum_{i \in G} C_{cap,i} P_{r,i} CRF_i(r, n_i) \tag{6.8}$$

$$CRF_i(r, n_i) = \frac{r(1+r)^{n_i}}{r(1+r)^{n_i} - 1}, \quad \forall i \in G \tag{6.9}$$

$$C_{ope}(t) = \sum_{i \in G} (C_{gas,i} f_{fuel,i}(t) + C_{m,i} P_i(t) + C_{su,i} SU_i(t)), \quad \forall t \in T \tag{6.10}$$

$$C_{pur}(t) = C_{buy}(t) \left[\frac{H_{elec\ heater}(t)}{\eta_{elec.\ heater}} + P_{grid+}(t) \right] - C_{sell}(t) P_{grid-}(t)$$
$$+ H_{bo}(t) \left[\frac{C_{gas,NG}}{u_{NG}\eta_{bo}} + C_{m,bo} \right], \quad \forall t \in T \tag{6.11}$$

$$E_{grid}(t) = K_{grid} P_{grid+}(t), \quad \forall t \in T \tag{6.12}$$

$$E_{DER}(t) = \sum_{i \in G} E_i(t), \quad \forall t \in T \tag{6.13}$$

$$E_{bo}(t) = K_{bo} H_{bo}(t), \quad \forall t \in T \tag{6.14}$$

where OF_1(cost) and OF_2(emission) are the objectives required to be minimized; f_{cap} is the capital cost of DERs; C_{ope} is the operational cost; C_{pur} is the energy purchase cost; E_{grid} is the CO_2 emissions from the main grid; E_{DER} is the total CO_2 emissions from all DERs in the microgrid; E_{bo} is the CO_2 emissions from the boiler; $C_{cap,i}$ is capital cost of installing DER i in \$/kW; CRF_i is the capital recovery factor of DER i; r is the interest rate; n_i is the lifetime of DER i in years; $C_{gas,i}$ is the gas price required for DER number i in \$/kg; $C_{m,i}$ is the maintenance cost of DER number i; $C_{su,i}$ is the start-up cost of DER number i; SU is the start-up status of DER number i at hour t; C_{buy}/C_{sell} is the buying/selling price of electricity from/to the main grid at hour t; P_{grid+}/P_{grid-} is the power bought from/sold to the main grid at hour t; H_{bo} is the heat supplied by the boiler at hour t; η_{bo} is the efficiency of boiler; $C_{m,bo}$ is the maintenance cost of boiler; K_{grid} is the carbon footprint for the energy purchased from the grid in kg CO_2/kWh; and K_{bo} is the emission from the boiler in kg CO_2/kWh.

b. *Constraints:*

$$\sum_{i \in G} P_i(t) + P_{grid+}(t) - P_{grid-}(t) = P_{ld}(t) + P_{losses}(t), \quad \forall t \in T \tag{6.15}$$

$$0 \leq \frac{H_{elec\ heater}(t)}{\eta_{elec\ heater}} + P_{grid+}(t), \quad P_{grid-}(t) \leq P_{grid,max}, \quad \forall t \in T \tag{6.16}$$

$$P_{i,\min} \leq P_i(t) \leq P_{i,\max}, \quad \forall i \in G, \quad t \in T \tag{6.17}$$

$$\sum_{i \in G} H_i(t) + H_{bo}(t) + H_{elec\ heater}(t) \geq H_{ld}(t), \quad \forall t \in T \tag{6.18}$$

where P_{ld} and H_{ld} are the power and heat demand at hour t; $P_{grid,max}$ is the upper capacity of the main grid; $P_{i,\min}$ and $P_{i,\max}$ are the lower and upper power generation of DER i, respectively; and $H_{elec\ heater}$ is the heat supplied by the electrical heater at hour t.

6.2.2 The Operational Stage

The proposed GP-EE schemes are installed and selected according to the terminal where it will be inserted. Proposed design (A) of (GP-EE) will be connected to the AC side, while proposed design (B) of (GP-EE) will be connected to the DC side. The global benefits from both of them will be enhancement of the ripple and harmonics in the voltage profile reducing power loss and enhancing the power factor. The detailed parameters of the microgrid, loads, DGs, and D-FACTS will be given later in the following sections.

The proposed GP-EE devices have two main schemes, as shown in Figures 6.3 and 6.4, based mainly on a combination of capacitors, filters connections, and MOSFET switches.

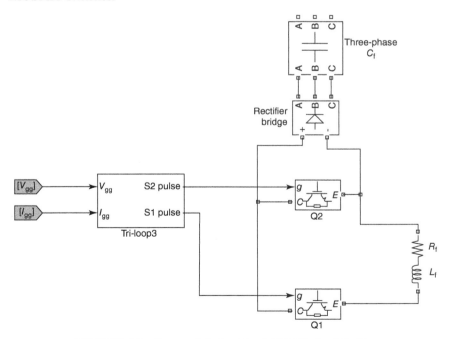

FIGURE 6.3 Proposed design (a) of (GP-EE) for AC side.

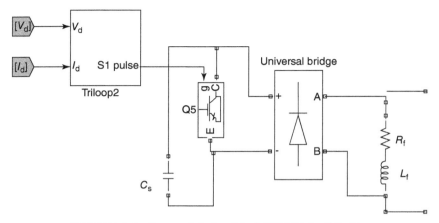

FIGURE 6.4 Proposed design (b) of (GP-EE) for DC side.

Typically, a hybrid combination of shunt capacitors with a tuned arm power filter constitutes the main GP-EE scheme (A) structure. Shunt capacitor banks are connected in parallel, where they provide reactive power compensation and improve the regulation level [11]. The capacitor works for dynamic voltage boosting and transient currents limiting. The path of solid-state switches has six-pulse converter-bridge plus resistance (R_f) and inductance (L_f) branch that structures a tuned arm filter. The main structure of universal bridge rectifier has six solid-state switches (six-pulse converter-bridge) where it receives from three-phase source and delivers to two-wire load, where V_{gg} and I_{gg} are the terminal voltage and current signals at GP-EE device. The two MOSFET switches (S1 and S2) are controlled by two complementary switching pulses (P1 and P2) that are supplied by the dynamic triloop error-driven modified dynamic PWM controller, which will be described later. First pulse P1 directs S1, while the second pulse P2 directs S2. Procedure of the complementary PWM pulses can be explained as follows:

Case 1: If P1 is high and P2 is low, the resistor and inductor will be fully short-circuited and the device will provide the required compensation to the load.

Case 2: If P1 is low and P2 is high, the resistor and inductor will be connected into the circuit as a tuned arm filter.

GP-EE scheme (B) structure depends on switchable capacitor behind universal bridge with resistance (R_f) and inductance (L_f) branch, where V_d and I_d are the terminal voltage and current signals at GP-EE device. MOSFET controller is fed by errors signal of PWM module that generates switching pulses to filter switch.

MEG configuration has different AC and DC distributed generating (DG) units that supply different AC and DC sides. Based on the output of the planning stage, the AC sources are DFIG wind turbine generator and microgas turbine generator. The DC sources are battery, fuel cell stack based on hydrogen, and PV arrays. For full utilization operation, there are boost converters; AC/DC, DC/AC, and DC/DC converters are connected to the microgrid. Various types of DC and AC loads can be represented, such as resistive loads, motorized DC series motor loads, linear AC loads, nonlinear AC loads, and three-phase induction motorized loads.

6.3 ENHANCED DYNAMIC PID CONTROL

The proposed modulated filter-capacitor compensation scheme is governed by adapting the PWM pulsing signal switching of GP-EE by selecting dynamic optimal gain values that dynamically reduce the global dynamic error. The resultant of error signal will initiate the enhanced dynamic controller according to the triloop-driven error that is firing to feed the control unit of PWM switching circuit, which will depend on the MOSFET gate to control the modified converter controller as shown in Figure 6.5. There will be GP-EE devices with four variables, both V_S and I_S are the composite voltage and current of the source bus of GP-EE and both V_L and I_L are the composite voltage and current of the load bus of GP-EE. The global error (e1) for the first GP-EE is the resultant of the six-loop individual errors, including the following:

1. V_L: Load bus voltage stabilization loop by tracking the error of the load voltage and regulating the voltage to near unity.
2. I_L: Dynamic RMS-current minimization loop to compensate any sudden current changes or oscillations.
3. P_L: Excursion dynamic damping loop for power oscillations.
4. Load ripple and transient current damping loop to reduce the harmonic ripple content in the system.

There will also be two novel extra supplementary loops to ensure the source power factor correction and the oscillations in the source current. The main per-unit three-dimensional error signals are $e_{V_L}, e_{I_L}, e_{P_L}$, and the other error signals are derivatives from them; these main error signals are governed by the following equations:

$$\text{error signal} : e_{V_L} = V_{L\,\text{ref}_{p.u.}} - V_{L_{p.u.}}\left(\frac{1}{1 + ST_1}\right) \tag{6.19}$$

$$\text{error signal} : e_{I_L} = \left(I_{L_{p.u.}} - I_{L_{p.u.}}\left(\frac{1}{1 + ST_2}\right)\right) \tag{6.20}$$

$$\text{error signal} : e_{P_L} = I_{L_{p.u.}} \cdot V_{L_{p.u.}}\left(\frac{1}{1 + ST_3}\right) \cdot \left(1 - \frac{1}{S + \alpha}\right) \tag{6.21}$$

$$\text{The global error(e1)} = \sum_{i=1}^{6} \beta_i \cdot \text{error signal}_i \tag{6.22}$$

The global error (e2) for second GP-EE and the global error (e3) for third GP-EE are calculated in the same way as e1. The optimization search algorithm is

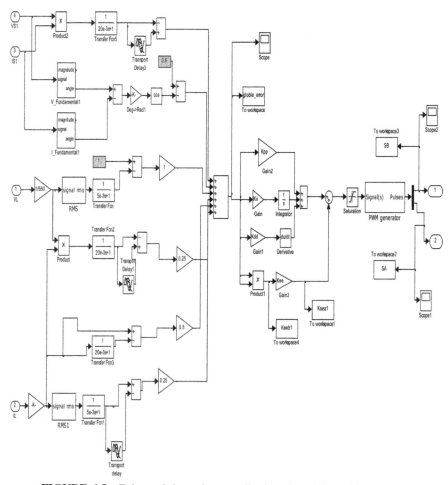

FIGURE 6.5   Enhanced dynamic controller based on triloop-driven error.

implemented for tuning PID controller gains (K_p, K_i, and K_d), for the three D-FACTS devices to minimize the system objective function. The proposed objective is defined by the global errors of all controllers:

$$OF = \text{minimization}(e1 + e2 + e3) \qquad (6.23)$$

6.4 BACKTRACKING SEARCH ALGORITHM

Resultant global error is the summation of individual multierror loops; each multidynamic leading loop is used to minimize the global error based on a triloop functional error signal in addition to other supplementary power factor correction and/or feeder currents limiting. To enhance system dynamic response, back-tracking search optimization algorithm will be applied to control the parameter settings of PID sections and to fine-tune the system dynamic response. BSA is used to control self-regulated PID triloop stage for GP-EE device. Proposed technique is used to accomplish a better efficiency where it can realize power quality improvement. Self-tuned PID controller adjusts the switching duty cycle ratio based on BSA technique. Output of the PWM generator is a train of pulses with variable duty cycles and constant frequency. BSA is a heuristic algorithm working on global minimum based on iterative population. There are many superiority points in applying BSA. First of all, the structure of BSA is quite simple; thus, it is easily applied on various optimization problems. BSA's mutation process uses only one individual from a previous population, and BSA's crossover process is more complex than in others. The crossover process in BSA is different than in the other heuristic techniques, the complexity means here it is advanced. In BSA, crossover process defines the final trial population with better optimized fitness values, while initial trial population is defined by mutation process. BSA's crossover process has multisteps to allow and control the number and selection of individuals that will mutate in a next trial. Also, BSA's crossover process has the option of ability to make individuals of the trial population overflow through allowed search space limits. BSA remembers the population of a randomly selected generation for use in calculating the search-direction matrix. Also, BSA is a dual-population algorithm that uses both the current and historical populations. Then, BSA's technique has a parameter that controls the trends and directions for searching, and can be adjusted to be large value to work on global search or small value to work on local search. Finally, BSA's boundary control mechanism is a very efficient tool for realizing the population diversity.

The procedure of BSA can be summarized in five main processes: initialization, selection-I, mutation, crossover, and selection-II, as described in the following, and as per Figure 6.6.

(A) *Initialization*
   The population P of BSA is initialized according to the following equation:

$$P_{ij} \cong (\text{Lower}, \text{Upper})_j, \quad \text{where} \quad i = 1, 2, 3, \ldots, N \quad \text{and} \quad j = 1, 2, 3, \ldots, D \tag{6.24}$$

N and D are population size and problem dimension, respectively, and U is the uniform distribution.

(B) *Selection-I*
   This step calculates historical population P_{old} to be used for determining search direction by applying following formula on pervious iterations:

$$P_{ij_{\text{old}}} \cong U(\text{Lower}, \text{Upper})_j, \quad \forall \text{ previous iterations} \tag{6.25}$$

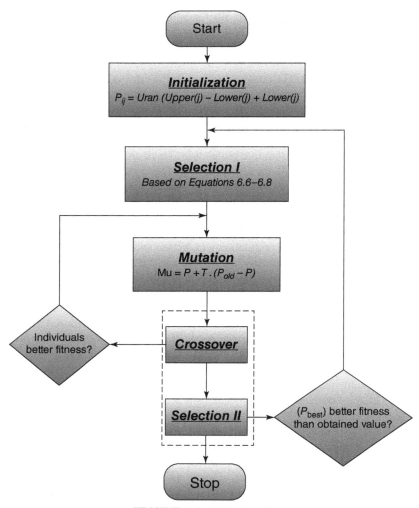

FIGURE 6.6 BSA flowchart.

At the beginning of a new iteration, the technique redefines historical population P_{old} as follows:

$$a < b, \quad \text{then} \quad P_{old} : P|a, b \sim U(0, 1) \qquad (6.26)$$

Here a, b are fitness values. Thus, BSA has a memory that remembers this historical population until it is changed. After determining P_{old}, a change in the order of the individuals in P_{old} is made by a random shuffling function:

$$P_{old} := \text{permuting}(P_{old}) \qquad (6.27)$$

(C) *Mutation*

BSA's mutation action presents initial frame of the trial population mutant:

$$\text{Mu} = P + \tau \cdot (P_{old} - P) \tag{6.28}$$

Here τ is a parameter that controls the trends and directions for searching; it can be adjusted to be large value to work on global search, or be small value to work on local search. BSA uses the historical population to take advantage of the experiences from previous generations.

(D) *Crossover*

Final population T is produced in this step. The initial population is Mu, as set in the mutation process. Individuals, which have better fitness values during the optimization process, are used to evolve final population individuals. BSA's crossover process has two steps. First one determines a binary integer-valued matrix (map) of size $N \cdot D$ that shows the individuals of T to be processed and updated by P. BSA's crossover strategy has efficient options different from the crossover strategies used in other heuristic algorithms. First option is based on mixed rate parameter in BSA's crossover action to control number of elements of individuals that will mutate in a trial. Second option allows only one randomly chosen individual to mutate in each trial and give it the ability to overflow the allowed search space limits as a result of BSA's mutation strategy.

(E) *Selection-II*

In BSA's Selection-II stage, a greedy selection is used to update the population with better fitness values than the others. The global minimizer is updated according to P_{best}, and the global minimum value is updated to be the fitness value of P_{best}.

6.5 CASE STUDY AND SIMULATION RESULTS

The digital simulations, MATLAB-Simulink-SimPower platform, with support of hardware demonstrations will concern with MEG Study System as the following specifications:

- *Utility Grid:* 138 kV, 5 GVA, $X/R = 10$
- *Micro Gas Turbine:* $V = 1.6$ kV, $P = 2000$ kW
- *Wind Turbine Generator:* $V = 1.6$ kV, $P = 1$ MW
- *PV:* 240 V, 500 kW, $N_s = 318$, $N_p = 150$, $T_x = 293$, $S_x = 100$, $I_{ph} = 5$, $T_c = 20$, $S_c = 205$
- *Fuel Cell:* 240 V, 2000 kW, number of cells $= 220$, nominal efficiency, 55%
- *Battery:* 240 V, rated capacity: 300 Ah, initial state-of-charge: 100%, discharge current: 10, 5 A

- *Hybrid AC Load 1:* linear load: 0.1 MVA, 0.8 lag pf, nonlinear load: 0.2 MVA, motorized load is an induction motor: three- phase, 0.3 MVA, 0.85 pf.
- *Hybrid AC Load 2:* linear load: 200 kVA, 0.8 lag pf, nonlinear load: 200 kVA, motorized load is an induction motor: three-phase, 100 kVA, 0.8 pf
- *DC Load:* resistive load: 100 kW, motorized load DC series motor: 100 kW
- *GP-EE(A):* $C_{f1} = 50\,\mu F$, $R_{f1} = 0.25\,\Omega$, $L_{f1} = 3\,mH$
- *GP-EE(B):* $C_s = 100\,\mu F$, $R_s = 0.1\,\Omega$, $L_s = 10\,mH$

The first stage is a planning stage, which concerns with an optimal deployment with respect to capacity sizes and types of DERs and CHP systems within microgrids. The objective is to simultaneously minimize the total net present cost and carbon dioxide emission. A multiobjective GA is proposed in that stage to handle the planning problem, including the optimization of DER type and capacity. The constraints involve power and heat demand constraints. Candidate DER technologies are CHPs with various characteristics, boilers, renewable generators (wind and photovoltaic), and a main power grid connection. Costs of CHP generators are based on their types and the capacity range.

6.5.1.1 The Planning Stage
The load profile for a year is divided into four seasons; each season was represented by 1 day, with total equal to $4 \times 24 = 96$ h. Then, the total cost (capital, operational, and maintenance), and CO_2 emission are calculated. It will change based on daily load profile. Table 6.1 summarizes the results of optimal planning.

The results of that stage will be fed to the second operational stage to ensure that it works on the optimal patterns of the DERs. The second stage concerns with the operational control aspect. GP-EE has been validated for enhancing the performance and power quality aspects of the MEG with DERs.

6.5.1.2 The Operational Stage
A novel control adaptation for error-driven dynamic multiloop regulator of proposed GP-EE by BSA technique has been applied to improve the power quality and energy utilization. The proposed GP-EE schemes with controlled BSA are very effective in enhancing power quality, improving power factor, reducing feeder losses and stabilizing the buses voltage,

TABLE 6.1 Optimal Planning Result

Case	NGT (kW)	H₂FC (kW)	NGFC (kW)	WT (kW)	PV (kW)	Total Cost ($)	Total CO_2 Emission (kg)
Min cost only	1048	13	0	27	102	1.368×10^5	8380
Min emission only	391	1300	0	90	6	7.231×10^5	4306.94
Compromised	1179	751	0	35	406	6.511×10^5	4876.4

TABLE 6.2   Values of Optimized PID Controller Gains

	Optimal Values of PID Controller Gains		
	K_p	K_i	K_d
GP-EE(A) 1	12	2.4	0.5
GP-EE(A) 2	15	5	0.2
GP-EE(B)	11	1.7	0.2

and mitigating the harmonic distortion. BSA system is applied to get the optimal PID values of GP-EEs as in Table 6.2.

The digital simulation is carried out with and without the controlled GP-EE, located in order to show its performance in voltage stabilization, harmonic reduction, and reactive power compensation at normal operating condition. Also, reduction of the enhancement in the power factor and managing the exchange power between MG and utility grid are all achieved and shown in Figures 6.7–6.13. The dynamic responses of voltage, current, active power, reactive power, apparent power, power factor, frequency spectrum (for voltage and current), (THD)$_v$, and (THD)$_i$ at source buses and load buses with comparison of harmonics at each bus are made in case of with and without GP-EE (Table 6.3). Voltage and current harmonic analysis in terms of total harmonic distortion (THD) is given. It is obvious that the voltage harmonics are significantly reduced, and the THD of current waveform at each bus is decreased.

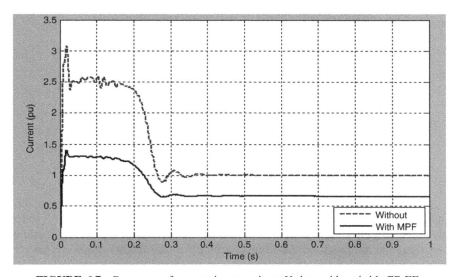

FIGURE 6.7   Response of current in per unit, at V_S bus without/with GP-EE.

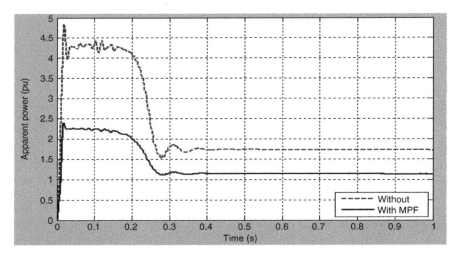

FIGURE 6.8 Response of complex power in per unit, at V_S bus without/with GP-EE.

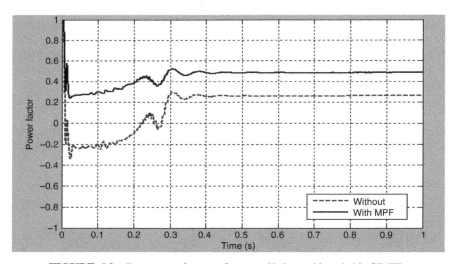

FIGURE 6.9 Response of power factor at V_S bus without/with GP-EE.

From all the previous figures, it can be observed that controlled GP-EE mitigates the harmonic and can achieve other technical benefits. Comparing the dynamic response results without and with the proposed GP-EE, it is quite apparent that the proposed GP-EE enhanced the power quality, improved power factor, compensated the reactive power, and stabilized the bus voltages.

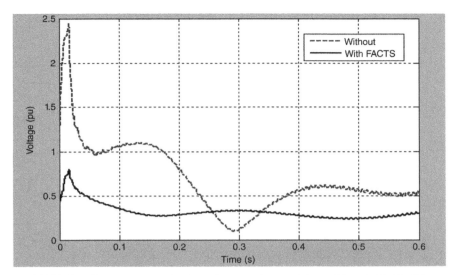

FIGURE 6.10   Response of voltage in per unit, at AC bus without/with GP-EE.

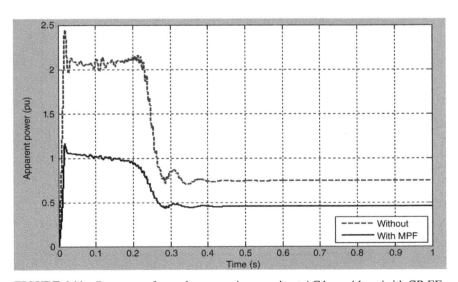

FIGURE 6.11   Response of complex power in per unit, at AC bus without/with GP-EE.

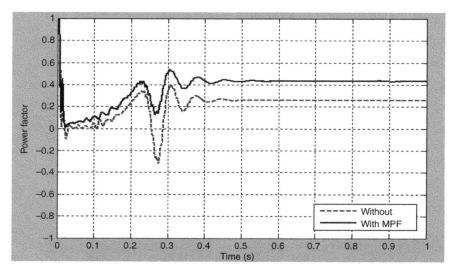

FIGURE 6.12   Response of power factor at AC bus without/with GP-EE.

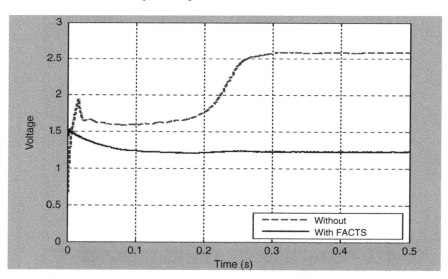

FIGURE 6.13   Response of voltage in per unit at DC bus without/with GP-EE.

TABLE 6.3   Percentage THD of Voltage and Current at the Buses

	V_S		V_1		V_L		V_g	
	THD$_v$	THD$_i$	THD$_v$	THD$_i$	THD$_v$	THD$_i$	THD$_v$	THDi
Without D-FACTS	0.62	7.25	35.5	21	29.3	36.8	26.7	18.5
With D-FACTS	0.1	4.55	4.82	4.56	4.4	4.92	4.61	4.2

6.6 CONCLUSIONS

This chapter presents a design of MEG with hardware demonstrations in ESCL, University of Ontario Institute of Technology. DERs are widely used in MEGs to match the various load types and profiles. DERs include solar PV cells, wind energy, fuel cell, battery, and microgas engine and storage elements. MEG will have both DERs and gas power sources as CHP, supplying thermal and electrical loads. MEG will include AC/DC circuits, FACTS, inverters, and power electronics controllers. Novel D-FACTS device, GP-EE with two DC/AC schemes, will be installed in MEG design. BSA can be applied to control the parameter settings of GP-EE to fine-tune the system dynamic response. BSA will adapt the dynamic gains of PID regulation triloop. The proposed controller ensures the adaptation of the global control error to operate with online optimal profile. The proposed strategy leads to get full MEG utilization by increasing the energy efficiency and reliability that ensure various technical impacts. Power factor improvement, bus voltage stabilizing, feeder loss minimization, and power quality enhancement are realized and achieved. The chapter concerns with two stages: planning platform and operational platform. The planning stage concerns with collecting load profile for 1 year, 4 seasons, for both thermal and electrical loading conditions and to optimize the size of each power source type based on cost and emission. The operational stage will evaluate and enhance the dynamic response, based on the assigned rates from the planning stage to the fine-tuning of the dynamic response by D-FACTs based on BSA. Hardware demonstration with digital simulations have validated the results to show the effectiveness and improved performance using the proposed strategy.

REFERENCES

1. Wu, X. and Wang, X. (2014) A hierarchical framework for generation scheduling of microgrids. *IEEE Transactions on Power Delivery*, 29, 2448–2457.
2. Guo, L., Liu, W., Cai, J., Hong, B., and Wang, C. (2013) A two-stage optimal planning and design method for combined cooling, heat and power microgrid system. *Energy Conversion and Management*, 74, 433–445.
3. Basu, A.K., Bhattacharya, A., Chowdhury, S., and Chowdhury, S.P. (2012) Planned scheduling for economic power sharing in a CHP-based micro-grid. *IEEE Transactions on Power Systems*, 27, 30–38.
4. Zhang, D., Evangelisti, S., Lettieri, P., and Papageorgiou, L.G. (2015) Optimal design of CHP-based microgrids: multiobjective optimisation and life cycle assessment. *Energy*, 85, 181–193.
5. Comodi, G., Giantomassi, A., Severini, M., Squartini, S., Ferracuti, F., and Fonti, A. (2015) Multi-apartment residential microgrid with electrical and thermal storage devices: experimental analysis and simulation of energy management strategies. *Applied Energy*, 137, 854–866.

6. Lund, H., Andersen, A.N., Østergaard, P.A., Mathiesen, B.V., and Connolly, D. (2012) From electricity smart grids to smart energy systems: a market operation based approach and understanding. *Energy*, 42, 96–102.

7. Andersen, A.N. and Lund, H. (2007) New CHP partnerships offering balancing of fluctuating renewable electricity productions. *Journal of Cleaner Production*, 15, 288–293.

8. Basu, A.K., Chowdhury, S., and Chowdhury, S.P. (2010) Impact of strategic deployment of CHP-based DERs on microgrid reliability. *IEEE Transactions on Power Delivery*, 25, 1697–1705.

9. Lian, K., Zhang, C., Gao, L., and Shao, X. (2012) A modified colonial competitive algorithm for the mixed-model u-line balancing and sequencing problem. *International Journal of Production Research*, 50 (18), 5117–5131.

10. Karami, S. and Shokouhi, S.B. (2012) Application of imperialist competitive algorithm for automated classification of remote sensing images. *International Journal of Computer Theory and Engineering*, 4, 137–143.

11. Lund, H. and Salgi, G. (2009) The role of compressed air energy storage (CAES) in future sustainable energy systems. *Energy Conversion and Management*, 50, 1172–1179.

CHAPTER 7

PERSPECTIVES OF DEMAND-SIDE MANAGEMENT UNDER SMART GRID CONCEPT

ONUR ELMA[1,2] and HOSSAM A. GABBAR[2]
[1]Department of Electrical Engineering, Yildiz Technical University, Istanbul, Turkey
[2]Faculty of Energy Systems and Nuclear Science, University of Ontario Institute of Technology, Oshawa, Canada

7.1 INTRODUCTION

The requirement of electrical energy by humankind has been increasingly continuing since the mid-1880s because of its key role in basic modern life, as well as the growing globalization with industrialization and information society in the world. Based on the rising energy requirement, the need to control demand-side management (DSM) of existing electrical resources efficiently emerged. In that respect, today an optimized energy management solution that promotes eco-friendly use of energy is very important. The dissemination of this solution and compatible work with the grid can be provided by transforming the traditional power grid into smart grid (SG) where information technology is integrated to traditional grid structure, so that every moment of the power generation and power demand can be monitored and controlled [1,2].

Demand-side management has a key role in the smart grid concept: to control the demand-side consumption and reduce peak loads. The ability to shift peak loads and provide the energy efficiency through better demand-side management is currently one of the most promising approaches to solve problems related to peak demand. DSM has some different terms such as demand-side energy management (DSEM), load energy management (LEM), demand response (DR), and automated

Energy Conservation in Residential, Commercial, and Industrial Facilities, First Edition.
Edited by Hossam A. Gabbar.
© 2018 The Institute of Electrical and Electronics Engineers, Inc. Published 2018 by John Wiley & Sons, Inc.

load management (ALM); all terms refer to the balancing of energy generation and consumption [3]. DSM includes all processes in demand energy systems such as utility renovation operations, metering, energy pricing, monitoring, customer comfort, home energy management systems, and so on. Hence, for DSM to be prevalent with its complete integration requires advanced communication technologies (ACT), intelligent devices, sensors, advanced metering infrastructure (AMI), and advanced algorithms. DSM is used for all type loads: industrial, commercial, and residential. However, most of the studies are focused on consumptions of buildings (commercial and especially residential loads). One of the reasons of this is that electrical energy consumption by residential and commercial buildings makes up a large percentage of the total electricity generated. For instance, in United States, commercial buildings utilize 54% of the total electrical energy and residential buildings consume approximately 40% [4,5]. Similarly, in Europe, residential and commercial buildings consume 50% of the total electricity consumption [6]. Another reason is that the concepts of environmental awareness and energy efficiency are intended to spread to all individuals in society. Therefore, many countries fund a lot of studies and projects in this regard.

Most of the developed and developing countries have some policies for increasing conservation and energy efficiency. These policies directly affect the customers in getting more benefits from the applications. Also, they set DSM according to peak demand loads and meet the generation needs with renewable resources under distributed generation concept. Because of increasing energy demands, governments tend to move to lower carbon emission energy sources (alternative energy resources). Also, DSM has the advantage that it is cheaper to wisely influence loads than to install new power plants or some electric storage systems.

All these DSM strategies are used for optimum and efficient energy consumption. So we can correlate between DSM and energy conservation such as in Figure 7.1.

Actually, all the demand-side energy management process is connected with each other and all of them are intended to provide informed energy consumption. They can be collected under progression of the DSM.

DSM is not a new method for electrical energy systems. The term DSM was first invented by Clark Gellings, who is a member of Electric Power Research Institute (EPRI), in 1984 and it was known as load management at that time [7]. During that time period, DSM was a widespread application area, especially with Smart Grid after 2000s. Various DR response programs, tariffs, and efficiency applications are already served by utilities [8–10]. Many demonstration projects and case studies are organized and analyzed by potential results of DSM applications [11–14]. In one of them, the DSM is integrated to a commercial building at Adobe Towers in San Jose, California; they have achieved 23% saving in energy consumption and reduced their total demand by 39% [15]. Also, several studies have been carried out related to DSM in the literature. Some studies focus on interaction between each end user and utility company [16,17]. Also, a different framework using DSM

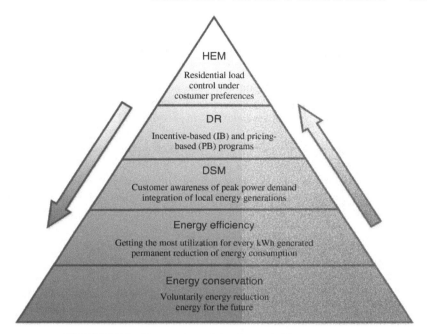

FIGURE 7.1   Progression of the demand-side management.

capabilities is suggested for customers with high electricity consumption [18]. In another study, the price-based optimization technique is investigated to maximize aggregator profit to control DR sources [19]. In addition, as an example of residential demand control, a home energy management algorithm is implemented using DSM technique to minimize the cost [20].

In the following sections some details of the DSM is explained and some benefits about DSM implementations is given. Also, in the following section, the DR techniques are explained, DR programs are classified, and DR application standards are given. Then, the impact of DR programs in residential energy management perspectives is discussed. Some details about HEM concept and examples are given. In conclusion, the existing DSM applications and what could be done in the future works are discussed.

7.2 DESCRIPTION OF THE DEMAND-SIDE MANAGEMENT

DSM can be defined in several different ways. However, to define comprehensively, DSM refers to initiatives and technologies to optimize energy use and involves programs or strategies that assist energy efficiency, conservation, and more effective energy management of electric consumption. On the other hand,

DSM and DR are used interchangeably. But, demand-side management and demand response are not exactly same methodologies, which are used interchangeably. DR is a particular form of the DSM focusing on load shifting features and moreover it aims to make customers efficient energy users in the long term. Thus, DSM architecture is designed to influence the period of use of electricity, so end users can optimally manage electricity usage. It can be divided into four classes for a better understanding of implementations in this regard [21]:

- Pricing incentives or signal dependent, short-period energy and capacity provinces that are done by the customers willingly.
- Factors such as active company control and advanced scheduling lead to capacity savings. Once the consensus is done with the customers, depending on the defined limits and parameters, the reduction and timing of the load is held.
- The customers consume more managed and efficient energy with the sustainable energy resources via more technological equipment, lighting, smart appliances, or control units.
- Conscious customers' willing actions in order to decrease cost or benefit the environment through education, communication, and public requests results in energy and/or capacity reduction.

On the other hand, the proportion between the level of peak loads and the average load in the electrical grid is called a load factor. Generally, the load factor is preferred to be high because of more efficiency and less required generation and increased transmission/distribution capacities. The load factor is defined by the power demand properties of devices and/or systems and the energy consumption habits of the end users. Different applications of DSM are usually conducted to improve the load factor, as shown in Figure 7.2. These four basic categories are generally used by the demand-side power management [22].

In addition, DSM has a key role in the energy management of the future smart grid, which provides support toward smart grid feasibilities in various topics such as control, management, infrastructure construction, and management of decentralized energy resources. Managing and controlling energy demand can decrease

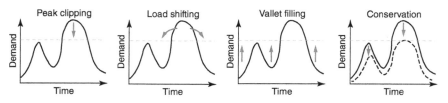

FIGURE 7.2   Categories of the demand-side management.

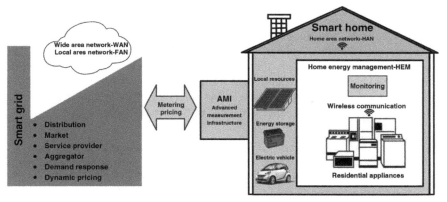

FIGURE 7.3   DSM framework architecture.

the overall peak load demand, restate the demand profile, and increase the grid sustainability by reducing carbon emissions. In this context, the efficiency that DSM provides in the electrical energy system depends on the architecture that it creates between generation and the end user as shown in Figure 7.3. Also, some of basic components of the DSM architecture are the following [23]:

- *Communication Systems:* Uses wired and/or wireless communication protocols to transfer all data bidirectionally.
- *Energy Management:* Engages in communication with the other units of the system and controls the electric consumption of end users based on the DSM system.
- *Local Generators:* Local energy generation units such as PV panels, wind turbines, and different types of units generating electric energy that can be used either locally or/and fed into the grid.
- *Smart Devices and Sensors:* Electric appliances are upgraded with intelligent hardware that provides monitoring, remote control, and management of the appliances. Also, sensors such as smart plugs used to monitor and control several data within the house.
- *Smart Metering:* Measures the end-user energy consumption and is equipped with AMI infrastructure. Smart meter can be integrated with smart devices and sensors to control these appliances.
- *Energy Storage Systems:* DSM system can be flexible to manage electric resources with energy storage systems.
- *Smart Grid Fields:* Especially, the distribution side domains can be included such as service providers, operations, aggregators, energy market, and costumers.

7.2.1 The Benefits of the DSM

In general, DSM benefits can be divided into two main parts: as benefits to consumers and benefits to the utilities. First, customers have the opportunity to decrease their electricity bills through energy efficiency measures with DSM. Second, the energy system can benefit from the shifting of energy consumption from peak to nonpeak hours and increase the efficiency of the system operation. In addition, these basic gains create a variety of benefits as follows [24]:

- *Energy Marketing Financial Benefits:* Demand response decreases the need for high-cost energy plants, which activates at high demand periods, so it provides low-cost energy in energy market. In the long term, because the demand response decreases the need of the wholesale markets capacity, it benefits the end user's energy bills.
- *Reliability Benefits:* Reliability of the power system can be raised by increasing operational efficiency and safety. This is also achieved with demand response by reducing the number of discontinuities that cause customers to experience various problems and financial losses.
- *Market Performance Benefits:* DR prevents energy suppliers from selling at prices above their energy production costs.
- *Participant Financial Benefits*: Participating customers in DR programs will benefit from participation, such as reduced bills and incentive payments.
- *Environmental and Social Benefits:* These refer to energy conservation and to increase in energy efficiency; the quality of the physical environment will be better, leading to reduced greenhouse gas emissions.
- *Customer Benefits:* These include reducing energy bills, increasing customer comfort, and improving customer service and products. Also, they can monitor and control your electric energy use. Home automation systems can do this automatically, which is called home energy management (HEM).

In addition, DSM also provides significant benefits for a balanced system operation. Particularly, DSM contributes to the balance between demand and supply, especially with load shifting as shown in Figure 7.4.

Also, DSM supports the dissemination of distributed generation. The local energy generation and balancing of the system can be easily controlled with DSM. Loads can be supplied by locally generated energy immediately when it is available and the transmission losses are reduced. Also, energy storage units such as battery storage, flywheel, vehicle-to-grid (V2G), and so on can be operated easily with DSM infrastructure. Distributed generation can supply electricity to the electric vehicle (EV) charge stations, which have recently been employed in electrical system, as shown given in Figure 7.5.

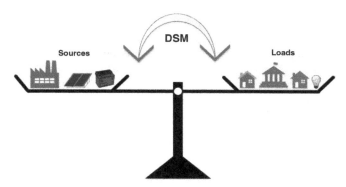

FIGURE 7.4   Role of the DSM in system stabilization.

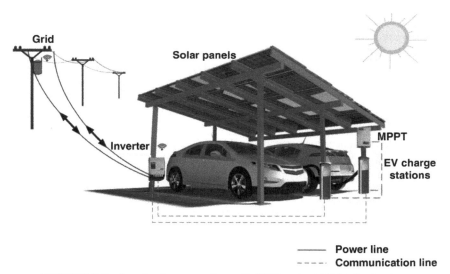

―――― **Power line**
‐ ‐ ‐ ‐ **Communication line**

FIGURE 7.5   Local solar generation with V2G under DSM infrastructure.

7.3 DEMAND RESPONSE

In electricity power system, demand can be influenced by various factors such as price, weather, technology, prosperity, population, and so on. Besides, DR involves reducing total consumption by affecting the consumers' consumption with various methods. To describe it in more detail, definition of the DR is given by Federal Energy Regulatory Commission (FERC) from the United States: "Changes in electric usage by end-use customers from their normal consumption patterns in response to changes in the price of electricity over time, or to incentive payments designed to induce lower electricity use at times of high wholesale market prices or when system reliability is jeopardized" [2,25]. DR provides flexibility to the electricity system and some other benefits such as reducing operating costs, loss reduction, stability and efficiency, and so on. In addition, DR is obtained through the application of DR resource types such as distributed generation, energy storage systems, and others that might contribute to alter the power supplied by the power system. Generally, DR uses mechanisms to induce consumers to decrease demand loads in order to restrict the peak loads; however, this might cause another peak load and limit the consumers' comfort. Therefore, DSM methods should be used to manage energy-use process.

In addition, direct utilities or some intermediaries assist end users to participate in DR programs. End-user demands are combined to obtain the aggregated end-user demand, which is used in the organized electricity markets. The aggregated end-user demand is provided by service providers (SPs), that is, an entity established for collecting such aggregated data from customers and/or demand response providers. The local distribution company can aggregate retail customers. Furthermore, they can manage the consumption to decrease peak load demand and/or optimize the electrical energy cost with adequate incentives for costumers they can coordinate [2].

7.3.1 Demand Response Programs

In order to influence customer behavior, the best option of DR approach is to achieve energy conservation. Also, DR programs can raise consumer awareness and contribute their ability using control and monitoring technologies. DR programs motivate the customers to show some positive effects of DR such as the aspiration to help avoiding blackouts, monetary savings, and a sense of responsibility [25]. In the other side of the DR, the utilities have a big responsibility to coordinate DR programs. The integrated set of coordinated services should be offered by the utilities to promote customers through different phases of program accession and technology adaptation. They should offer technological renovation and financial assistance such as locating and employing contractors. There are many versions of DR programs present depending on the way of application. Currently, various available DR programs that are summarized into some classifications are proposed in Table 7.1 [24]. We can observe two major categories of DR: incentive-based programs (IBPs) and price-based programs (PBPs) and their subtitles.

7.3.2 Examples of Demand Response Applications

The demand response is an important application to reduce peak shaving and to control energy consumption. Thus, several countries have some implementations of DR programs. Especially three of them are the United States, Europe, and China. The United States is the initiator country where DR applications

TABLE 7.1   Classification of the Demand Response Programs [24]

Price-Based DR	Incentive-Based DR
Time of Use (TOU): A rate having variable electricity prices with different slots of time (usually 24 h having equal time slot of 1 h). TOU pricing reflects the electricity unit price of the delivering power to the end users. TOU rates vary by time of a day (e.g., high peak, mid peak, and low peak). TOU tariffs are widely used by industrial and residential consumers *Real-Time Pricing (RTP):* A rate where electricity price fluctuates hourly depending on the changes in energy wholesale market. Generally, these prices are known in advance to customers a day ahead or hourly basis *Critical Peak Pricing (CPP):* It is designed in such a way that high price is predefined over certain intervals of time. CPP can be triggered due to high prices or by high system contingencies by utilities obtaining power from wholesale markets. CPP is not itself a pricing scheme. It can be superimposed to any other pricing scheme like RTP, TOU, DAP, and so on. Generally, it is implemented for limited time duration due to high energy demand or long-term system faults. CPP scheme is not very popular. However, participating customers get incentives during non-CPP hours	*Direct Load Control (DLC):* It is a program in which energy retailers or providers directly shut down some residential loads on short notices in order to address reliability contingencies. Utilities are given a full access to partial loads (e.g., pool pump, air-conditioners, and water heaters). Participating customers obtain incentives from retailers. In case of violating the agreement of participation, customers are imposed penalties. DLC programs are used for small industrial and residential customers *Interruptible/Curtailable Service (I/C):* These programs are integrated to ongoing customer tariffs and provide incentives or discounts in participating the load reduction during fault conditions. Generally, these programs are used in large industries or commercial sectors *Demand Bidding/Buyback:* In these programs (i) customers offer the utilities of the prices at which they agree to curtail their load, and (ii) encourage the end users how much load they would curtail on the given utility price. *Emergency DR:* Programs that offer incentives to the customers who are willing to participate in load reductions during fault conditions. Penalties for violating customers may or may not be used *Capacity Market Programs:* These are designed for those customers who are willing to curtail prespecified load during given time duration *Ancillary Services Market Programs:* These programs have a similar process like demand bidding. Some customers bid to limit their own loads. If their bids are accepted, they are paid market price because they are on hold. If load constraints are required, they can be paid at the energy price of the spot market

are most applied and studied. Most of the DR programs were implemented in the United States. In Europe, energy strategies are mostly focused on energy efficiency and DSM instead of DR applications. However, some of the DR programs discussed thus far focused on large industrial settings. China has some energy problems with the rapid development of social economy, the continuous upgrading of the industrial structure, and consumption structure. Thus, China implemented some DR pilot projects. Some DR application examples are given Table 7.2, especially from the United States, Europe, and China.

TABLE 7.2   Instances of Demand Response Applications [26]

	Measures
Applications in the United States	• *Time of Use:* 1.1 million residential customers participated in project of the time-sharing electricity in 2010. This project conduced rising of the demand response resource capacity, which is about 2.28 million kW every year in the United States
	Critical Peak Price: First, critical peak price program was applied at California State in 2011. It has limited the load to not less than 200 kW. During the application process, sometimes peak demand price increased approximately from \$24/kWh to \$224/kWh because the peak loads are aimed to limit and the operating complexity of the grid is reduced during the peak hours
	• *Real-Time Price:* Real-time pricing program was applied in California in the 1980s. This program is divided into two-parts: previous real-time pricing and real-time electricity pricing
	• *Direct Load Control:* Direct load control program was used by over 5.6 million customers in the United States in 2010. Direct load control project is provided about 9 million kW capacity every year in the United States. Generally, direct load control projects are regulating and controlling of electrical equipment, such as heater.
	• *Interruptible Load:* This program is suitable for large-size industrial and commercial users; annually, over 10.97 million kWh is obtained for demand response capacity
	• *Demand-Side Bidding:* Demand-side bidding is integrated into the process of day-ahead market. Users offer the utilities the prices at which they agree to curtail their loads. Also, customers can be encouraged to curtail their loads on the given utility price.

TABLE 7.2   *(Continued)*

	Measures
Applications in Europe	*Time of Use:* Some of the European countries offer some kind of TOU price rates: In Italy, especially in peak periods, from 8 a.m. to 7 p.m. in weekdays, the electricity price is increased. In France, different prices are shown with different colors marked in the peak times, flat times, and valley times *Interruptible Load:* In England, some of the processes are being used such as for frequency control and power system stabilization. In Finland, capacity building on the demand side to create frequency reserves through annual agreements signed with industrial users *Demand-Side Bidding:* In Norway, capacity market of the power system frequency modulation is developed; Demand-side program is used to achieve the power stabilization and frequency modulation
Applications in China	• *Time of Use:* This electricity price is extended to the provinces of Jiangsu, Zhejiang, and Shanghai for residential customers. For industry users, the tariff proportion is 5:1 in Jiangsu and Shanghai. In Yunnan electricity price decreases by 15% during wet period and also electricity price increases by 20% during dry season. The TOU price varies plus or minus 50% and was increased to the nonresidential and commercial electricity consumption of 100 kVA and above in wet and dry periods • *Critical Peak Price:* Critical peak price has been implemented in Shandong province from 2008. In Hebei province, more than 4000 enterprises implement the critical peak price and so more than 1 million kW capacity is obtained with peak load reduction

7.3.3   Information about Demand Response Standards

Development in the power system standards is an important necessity on the way to the transition to the smart grids. Also, the demand-side varieties have increased with the smart grids such as distributed generation, DSM applications, dynamic pricing, and DR programs. These different applications have been implemented in many countries and being used actively in the grid. Among these implementations, DR programs stands out more. The DR programs have various types and include many concepts such as shifting limits, notification times, response times, length of response, calculation methods, dynamic pricing, payment terms, and so on. On the other hand, DR opportunities need some factors to be applied such as energy markets, government policies, and local utilities. In addition, having defined standards between retail side and utilities for different DR programs benefits both. A standard demand response signaling helps the facilities in terms of management and load reduction calls and so more customers can be involved in DR products and programs. That is the reason why standardizing works are done in the United States especially for the DR programs. The most outstanding study in this field is the Open Automated Demand Response (OpenADR), which is being developed at Berkeley University Demand Response Center in the United States [27]. In this OpenADR system, with an open and industry-approved data model, a standardized SD communication and signaling infrastructure is served. The OpenADR has been continuing its development with its versions of OpenADR 1.1 and lately OpenADR 2.0. The proposal of the OpenADR 2.0 is to benefit both utilities and aggregators through DR communications. Finally, today the foundation and the rise of the OpenADR Alliance (www.openadr.org), put OpenADR to a wider market, which also serve certification, testing, training, and education [28]. The alliance is expected to raise the demand response and give more choices for the facility managers such as utility programs or products. The OpenADR is a pioneer study for the standardization of DR applications and continues its progression. Especially under the standardization heading, the importance of such movements is indisputable. These standards are very valuable requisites for the energy market, utilities, and customers. Particularly, the developed standards for DR programs will raise the customer confidence in markets because they will obtain more valid and objective measurement results.

7.4 SMART METERING

Smart meter (SM) has a key role in DSM and DR programs. The SM is considered the first and basic step for the smart grid in most of countries. SM allows implementation of advance metering infrastructure (AMI), two-way communication, and bidirectional power flow, which makes it more important for smart grid future vision. On the other hand, SM is one of the most important tools to provide opportunity to consumer and emerging the new market occasion [29]. According to recommendation of 2020 European Union (EU), installation of smart meter should be accelerated in order to accelerate the renewable energy and electrical vehicle integration to grid [9]. In addition, according to EU directions, it is stressed that smart meters contribute the energy efficiency and energy saving and SM installation rate is aimed at 80% in European Union until 2020 [30]. SMs measure electrical energy consumption and other billing parameters of any consumer and provide more information than traditional meters. SMs can also record these data and it can send the data back to a main server in predefined intervals for analysis, control, and monitoring. Features of the SMs are referred to in this section in detail.

Smart meters are implemented by developed and developing countries in their smart grid roadmaps. First step of SG implementation is using SM instead of

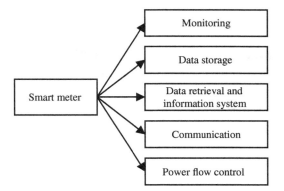

FIGURE 7.6 Typical smart meter features.

traditional electromechanical meter. These meters could measure instantaneous and billing parameter such as voltage, current, power, power factor, maximum demand, and load profile. Smart meters are low-cost commodity devices that record consumption of electrical energy with two-way communication technology. Typical smart meter features are shown in the Figure 7.6.

SM offers the user interface notification and that is the most significant feature of SM unlike the traditional one. Visual data are introduced to user by the interface. It is mentioned that electric consumption can be reduced up to 15% with only notification to user [31]. Periodic recording of data is stored in data storage unit for beneficial analysis such as load forecast. This also provides a more detailed comprehension of the event occurrence. SM communication networks generally comprise two types of communication network: the wide area network (WAN) and the local area network (LAN). In Ref. [32], detailed information about communication for SM is given.

Smart meters are of two main types: advanced metering infrastructure (AMI) and automatic meter reading (AMR). While AMR allows one-way communication, AMI offers a more advanced measurement infrastructure with two-way communication. Also, smart meters provide significant operational and efficient improvements and have the following advantages [33]:

- Remote meter reading
- Reduced electricity theft and so increased energy security
- Quick identification and rectification of faults; low overall power outage duration so improved outage management
- Improved voltage management
- Detect customers who have installed distributed generation
- More advanced billing and customer support operations
- Allow demand response applications such as direct load control and central air control

- Improved efficiency of power system due to the fact that it can perform fault analysis, power quality analysis, and demand control
- Minimize the time of operational decisions
- Scheduling inhibitive maintenance
- More accurate billing
- Manual energy meter reading process, which is a difficult, continuous, and an expensive job, is simplified with smart meter
- Encourage consumers to conserve their energy
- The geographical location and the necessary information of a potential fault can be obtained

In addition, SM has several benefits to improve electric energy system under smart grids concept. The benefits of smart meter investments can be examined under four headings: customer benefits, distribution company benefits, supplier company benefits, and electrical market benefits [34]. Benefits of smart meters are summarized in Table 7.3.

DR manages the demand and assists consumers in changing their electricity consumption pattern. DR programs improve the reliability of the electric grid by managing stress on grid during peak times. Since percentage reduction of peak demand is used as performance criteria in DR programs generally, SM is the necessary tool to measurement DR impact. SM plays an infrastructure role in DR programs mentioned previously. In DR, SM is used to illustrate a typical three-step DR event: overload, notification, client response [35]. Overload means exceeding the predefined threshold value of system peak load. Notification means smart meter triggers information when contingencies are at peak loads. Client response means taking set of actions to reduce consumption such as turning off noncritical load by service providers. Smart meters also provide indirect benefit in demand-side management [36]. For example, if the end users have any customer-based

TABLE 7.3 Benefits of the Smart Meters

Customer Benefits	Distributed Company Benefits	Supplied Company Benefits	Electric Market Benefits
Energy saving	Quality	Profitability	Reduce CO_2
Discount on the bill	Operational efficiency	Specific customer tariff	Reduce energy cost
Customer satisfaction	Reduce operation cost	Reduce incorrect billings	More competitive
Energy monitoring	Remote on/off	Reduce operation cost	Reduce system imbalances
Energy management	Reduce lost and leak	More accurate demand forecasts	Widespread distributed generation

technologies that respond to SM signal such as smart appliances in their home, then they have more options related to DR programs such as direct load control, real-time pricing. So we can say that the more smart meters are used, the more DR opportunity is served.

7.5 DYNAMIC PRICING

The first pricing of consumed electrical energy is made by electromechanical meters in the classic grid system. Accordingly, the electrical energy distributors can apply only one type of pricing. However, the development of metering technologies enabled different pricing options. Recently, various pricing methods have become possible with the help of the smart grids. The exchange of old electrical meters with smart meters allows the two-way communication between smart home and electrical utilities. Therefore, the smart meters have a key role at the DSM [37,38].

Nowadays, the demand response management systems, which came to light for a more controlled energy consumption, are branched into two: incentive-based demand response and priced-based demand response, as mentioned previously. Priced-based demand response depends on a different calculation of the price at different defined periods of the day. So, the demand response can be controlled indirectly. The used dynamic pricing method depends on three main pricing structures: time of use (ToU), real-time pricing (RTP), and critical peak pricing (CPP). The mentioned pricing systems are especially more compatible with smart home concept. A real-time pricing tariff is fixed by the electrical utility and transferred to the consumer via Internet or grid system. Besides, the real-time pricing examples from around the world are examined. One of these examples is the real-time pricing tariff that is put into practice by a distribution utility in the United States. In this application, both hour- and 5 min-dependent pricings are given at the Web site currently, as shown in Figure 7.7 [39].

7.6 RESIDENTIAL DEMAND CONTROL: HOME ENERGY MANAGEMENT

Residential loads have a significant rate in electrical energy consumption such that residential electricity consumption is nearly 37% of the total electricity consumption in the United States. On the other hand, according to report of International Energy Agency, electricity consumption increases about 3.4% every year in the world. Considering these rates, the residential electricity consumption should be managed for energy efficiency and costumer benefits. This management can be actualized by smart control system, which is called HEM or home energy management system (HEMS). HEM introduces a framework of the smart home

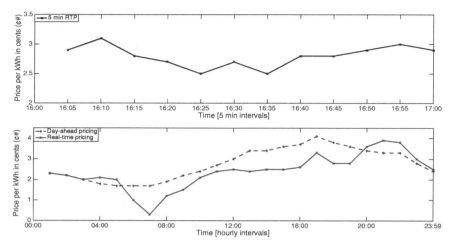

FIGURE 7.7 Examples of real-time pricing.

(SH) concept and has smart controller/controllable devices and advanced communication technologies.

Smart home has been applied since 1990s and is called home automation [40]. Home automation has been focusing on raising customer comfort. At first, the goal of the home automation system was to provide both comfort control and whole-house monitoring. But after the smart grid development, home automation concept was altered into smart home concept. Therefore, the terms energy efficiency, control, and comfort became more important in smart homes. In these homes, the consumers have the control to reduce/shift the power of their electrical appliances depending on the data taken from grid and the crucial role of home area networks (HAN) is to provide this communication between the grid and the appliances [41]. So, the end user makes the decision of maintaining energy efficiency and level of comfort. But HEMS makes the optimization of comfort and energy optimization. According to one study, only monitoring of real-time energy consumption can lead to approximately 25% reduction of energy demand, enabling end users to consume responsibly [42]. Also, HEM can reduce peak demand, save electricity costs, and meet the needs of DSM in smart grid. HEM has an important potential for demand response, peak-time rebates, critical peak pricing, time-of-use pricing, distributed generation, and electric vehicle integration. Thus, home energy management products are developed by many companies on the market.

The HEM can also work in coordination with the utility for considering energy efficiency and satisfying consumer comfort. Depending on DR signals and/or dynamic pricing coming from the utility, it can manage energy consumption at home. In addition, HEM systems can be integrated with local energy generation and also EV charging. In this case, HEM controls not only the loads but also

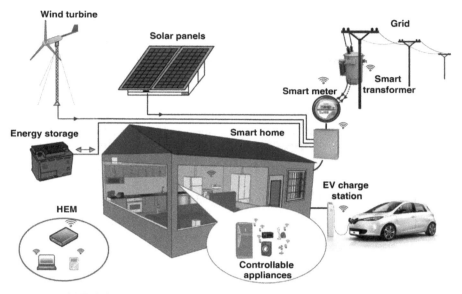

FIGURE 7.8   Advance HEM concept under smart home architecture.

generation. That is why HEM requires a more complex and advanced system such as that shown in Figure 7.8.

In smart home concept, both loads and local generation units compose a kind of microgrid structure that can be managed by HEM. This structure has a lot of different units. The thing to be considered is that the data that come from many different units should be transferred correctly and on time, which demonstrates the importance of communication system for HEM. There are some wired and wireless communication protocols that are used in HAN. Depending on the communications media, there are two kinds of communication technologies properly used in HAN applications. The first class makes connections through wires and technologies supported by Power Line Communications (PLC) and Ethernet. On the other side, the second class makes communication wirelessly, which are generally chosen as IEEE 802.15.4 (e.g., ZigBee) and Wi-Fi. The advantage of the first group is that these technologies can provide fast and more robust communication medium. However, wireless communications can provide low-cost infrastructure and ease of connection to unreachable areas. Nearly all of these protocols can provide point-to-point (P2P) communication called unicast as well as multicast communication in star, tree, and mesh networks. Also, several communication profiles based on IEEE 802.15.4 standard have been developed for smart home applications. Comparison of some communication protocols for smart home energy management systems are given in Table 7.4 [42].

IEEE 802.15.4 technical standard is formed to use in low-rate wireless personal area networks. Especially one of the communication protocols, ZigBee, is

TABLE 7.4 Features of Communication Protocols for HEMS

Communication Protocols	Standard	Data Rate	Coverage Distance	Frequency	Energy Consumption
Wi-Fi	IEEE 802.11n	72.5, 150 Mbps	100 m	2.4–5 GHz	Very high
ZigBee	IEEE 802.15.4g	20, 40, 250 kbps	75 m	868, 915 MHz 2.4 GHz	Medium
Z-Wave	ITU-T G.9959	9.6, 200 kbps	30 m	868, 915 MHz	Medium
En-Ocean	ISO/IEC 14543-3-10	120 kbps	30 m	315, 868 MHz	Very low
Dash7	ISO/IEC 18000-7-2004	27.7–200 kbps	100 m	433 MHz	Low
PLC	IEEE 1901.2	Up to 500 kbps	50 m	1–500 kHz	N/C

preferred because of advantages such as low cost, low data rate, short range, interference-resistant features, and low power consumption that is compatible with the ultralow duty cycle. Also, ZigBee communication protocol is designed to be more plain and cheaper than other wireless personal area networks such as Bluetooth or Wi-Fi. Thus, this wireless protocol is suitable for use in HEM which is IEEE 802.15.4 standard. There are three types of devices in ZigBee network. The first is a coordinator, which is used for coordination and forming to the network. Coordinator serves as a router in a mesh node. Only a network coordinator can be designed as a trust center. The second device is a router that makes the transfers between the network devices. It can also send and take messages on its own. The third one is an end device that makes the communication between the parent nodes only and is not able to send relay messages aimed at another node. Sleepy end devices both do not transfer direct messages and power off their radio. The structural and typical features of these devices are given in Figure 7.9.

In this respect, many pilot smart home projects using the technologies mentioned above are being carried out in different parts of the world. One of them is Honda smart home projects in the state of California [43]. This smart home has more than 200 sensors and also is supplied by solar energy system and has a EV charger for electric vehicle to provide zero emission living and mobility. Another example is Yildiz Technical University Smart Home (YTU-SH), which was built in the laboratory environment. The 35 m^2 prototype smart home consists of a living room with kitchen, a study room, and a bathroom. YTU-SH is equipped with residential appliances that are similar to those available in an average home. Unique smart home software was developed to monitor and control all appliances with smart plugs. This smart home is provided renewable resources such as solar and wind. Also, there is an EV charge station to charge EV with zero emission electrical energy produced by renewable resources [44]. These types of projects

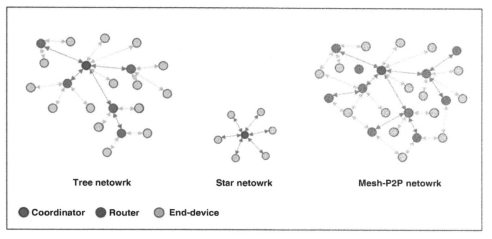

FIGURE 7.9   ZigBee communication devices and networks.

allow the possibility to detect challenges of implementations and help to provide solutions for these challenges. These projects fabricate a lot of valuable data that can be used for analysis of customers and utility side effects.

7.7 CONCLUSION

Electrical energy has become an indispensable part of human life and this valuable resource has to be used optimally. Thus, the energy management applications have a critical role in this. Energy management is called DSM under smart grid concept. In general, it has been recognized as an effective tool to reduce operational costs in industry and commerce and has some benefits for both customers and utility services. DSM has been implemented in most countries, mainly in the United States. There are several different types of applications under DSM such as energy efficiency, demand response, home energy management, smart metering, monitoring, and so on. All of these DSM applications have some challenges in application in the power system. Especially, DR programs have a variety of challenges. End users do not have a positive attitude for new pricing tariff and/or energy consumption limitation because of their habits of classic system. In addition, the necessary infrastructure is slowly progressed by utilities. It is also necessary to use the energy market more efficiently, educate customers about better technological equipment, and explain the benefits provided by these programs. For this reason, government incentives have demonstrated the positive outcomes of these programs. Also, as known DSM applications give a lot of opportunities, both customers and utilities enjoy environmental benefits.

One of the DSM applications is DR, which offers new perspectives on electrical energy generation and consumption and on the operation of power systems. It should also be noted that DR can also be applied to commercial and industrial customer levels. DR will play an important role in the SG implementations and will evoke benefits on system balance, operation, expansion, and efficiency. The integration of storage devices, distributed generation, and onsite RES in automated DR brings additional flexibility and complexity that should be managed with innovative technologies and methods. However, many more researches are needed to show all these positive effects. For example, electrical consumption by customers needs to be measured and analyzed. Also, heating, ventilating, and air-conditioning loads can effectively be managed to reduce the demand at peak hours for these customers. Since temperature change has long time constant, shifting these loads will only cause minimal discomfort. However, in DR applications, small time step, such as 5 min, should be used for maintaining a comfortable environment. As industrial and commercial customers consume high power loads, DR applications at these levels will possibly offer more benefits to the grid side for peak load management and economic operation. Also, DR will become a true and viable resource to introduce system players such as independent aggregators and retailers into the market. However, some countries such as the United States, China, and Europe will need to explain the effect of independent DR providers in order to real and measurable change through market competition. The standardized process allows for this competition around demand-side flexibility.

On the other hand, smart meter is one of the most important equipment in the DSM systems. Smart meters enable the demand-side management to be more flexible and provide different methods. Also, they can advise customers to use energy at suitable times, and on the other hand at grid side they provide a more economic operation and crucial advantages to follow/prevent leakages. SM can be integrated with HEM systems to manage energy consumption without compromising consumer comfort, and mediates to communicate between generators to end-use appliances.

All these DSM applications, programs, and equipment are being used to build future power system, Smart Grid. To achieve this purpose, it is obvious that there are much more researches need to be done. For example, one of these researches should aim to understand the optimal amount of customer versus demand response provider control of appliances. Another one can be an analysis of the environmental effects of demand response. Yet another study may be about how demand response resources can be dispatched to support and balance variable generation from renewable energy. Another important research can be about investigation of effects on appliances' durability, which are cycled on/off by demand response programs. It is possible to increase these study titles. However, today's initiatives and practices on demand control and energy efficiency are shedding light on and encouraging us to do so in the future. The DSM applications will be an indispensable part of the future power system.

REFERENCES

1. Palensky, P. and Dietrich, D. (2011) Demand Side Management: Demand Response, Intelligent Energy Systems, and Smart Loads. *IEEE Transactions on Industry Informatics*, 7, 381–388.

2. Siano, P. (2014) Demand response and smart grids: a survey. *Renewable and Sustainable Energy Reviews*, 30, 461–78.

3. Khan, A.A., Razzaq, S., Khan, A., and Fatima Khursheed, O. (2015) HEMSs and enabled demand response in electricity market: an overview. *Renewable and Sustainable Energy Reviews*, 42, 773–785.

4. Chua, K.J., Chou, S.K., Yang, W.M., and Yan, J. (2013) Achieving better energy-efficient air conditioning: a review of technologies and strategies. *Applied Energy*, 104, 87–104.

5. U.S. Department of Energy (2010) *Buildings Energy Data Book*. Available at http://BuildingsdatabookErenDoeGov/, 1–271.

6. Kok, K., Karnouskos, S., Nestle, D., Dimeas, A., Weidlich, A., Warmer, C. et al. (2009) Smart Houses for a Smart Grid. *International Conference and Exhibition on Electricity Distribution*, pp. 1–4. doi: 10.1049/cp.2009.0961.

7. Warren, P. (2014) A review of demand-side management policy in the UK. *Renewable and Sustainable Energy Reviews*, 29, 941–951.

8. Herter, K. and Wayland, S. (2010) Residential response to critical-peak pricing of electricity: California evidence. *Energy*, 35, 1561–1567.

9. Albadi, M.H. and El-Saadany, E.F. (2007) Demand response in electricity markets: an overview. *2007 IEEE Power Engineering Society General Meeting (PES)*, doi: 10.1109/PES.2007.385728.

10. Mohsenian-Rad, A.-H., Wong, V.W.S., Jatskevich, J., Schober, R., and Leon-Garcia, A. (2010) Autonomous demand side management based on game-theoretic energy consumption scheduling for the future smart grid. *IEEE Transactions on Smart Grid*, 1, 320–31.

11. Zeng, M., Xue, S., Ma, M., Li, L., Cheng, M., and Wang, Y. (2013) Historical review of demand side management in China: management content, operation mode, results assessment and relative incentives. *Renewable and Sustainable Energy Reviews*, 25, 470–482.

12. PG&E's demand response case studies. Available at https://www.pge.com/en_US/business/save-energy-money/energy-management-programs/demand-response-programs/case-studies/case-studies.page (accessed June 5, 2017).

13. Energex's Commercial and Industrial Energy Conservation and Demand Management Program. Available at https://www.energex.com.au/__data/assets/pdf_file/0004/342823/Goodman-Fielder-Case-Study.pdf (accessed June 1, 2017).

14. Cappers, P., MacDonald, J., Goldman, C., and Ma, O. (2013) An assessment of market and policy barriers for demand response providing ancillary services in U.S. electricity markets. *Energy Policy*, 62, 1031–1039.

15. PG&E's Integrated DSM Case Study in Commercial Building. Available at https://www.pge.com/includes/docs/pdfs/mybusiness/energysavingsrebates/

demandresponse/cs/OfficeBuildings_Adobe_Integrated_CaseStudy.pdf (accessed June 10, 2017).

16. Conejo, A.J., Morales, J.M., and Baringo, L. (2010) Real-time demand response model. *IEEE Transactions on Smart Grid*, 1, 236–242.

17. Pedrasa, M.A.A., Spooner, T.D., and MacGill, I.F. (2010) Coordinated scheduling of residential distributed energy resources to optimize smart home energy services. *IEEE Transactions on Smart Grid*, 1, 134–143.

18. Kazempour, S.J., Conejo, A.J., and Ruiz, C. (2015) Strategic bidding for a large consumer. *IEEE Transactions on Power Systems*, 30, 848–856.

19. Parvania, M., Fotuhi-Firuzabad, M., and Shahidehpour, M. (2013) Optimal demand response aggregation in wholesale electricity markets. *IEEE Transactions on Smart Grid*, 4, 1957–1965.

20. Nguyen, H.T., Nguyen, D.T., and Le, L.B. (2015) Energy management for households with solar assisted thermal load considering renewable energy and price uncertainty. *IEEE Transactions on Smart Grid*, 6, 301–314.

21. Demand, P. and Team, S.M. (2011) *2010 Annual Review of Energy Efficiency Programs: California.*

22. Wang, S., Xue, X., and Yan, C. (2014) Building power demand response methods toward smart grid. *HVAC&R Research*, 20, 665–687.

23. Barbato, A. and Capone, A. (2014) Optimization models and methods for demand-side management of residential users: a survey. *Energies*, 7, 5787–5824.

24. U.S. Department of Energy (2006) Benefits of demand response in electricity markets and recommendations for achieving them. doi: citeulike-article-id:10043893.

25. Shariatzadeh, F., Mandal, P., and Srivastava, A.K. (2015) Demand response for sustainable energy systems: a review, application and implementation strategy. *Renewable and Sustainable Energy Reviews*, 45, 343–350.

26. Dong, J., Xue, G., and Li, R. (2016) Demand response in China: regulations, pilot projects and recommendations – a review. *Renewable and Sustainable Energy Reviews*, 59, 13–27.

27. Holmberg, D. (2011) Demand response and standards new role for buildings in the smart grid. *ASHRAE J*, 53 (11), 23–28.

28. OpenADR. Available at http://www.openadr.org/ (accessed June 25, 2017).

29. Sharma, K. and Mohan Saini, L. (2015) Performance analysis of smart metering for smart grid: an overview. *Renewable and Sustainable Energy Reviews*, 49, 720–735.

30. Council of the European Union (2012) Commission Recommendation 2012/148/EU on Preparations for the Roll-Out of Smart Metering Systems.

31. Faruqui, A., Sergici, S., and Sharif, A. (2010) The impact of informational feedback on energy consumption: a survey of the experimental evidence. *Energy*, 35, 1598–1608.

32. Erlinghagen, S., Lichtensteiger, B., and Markard, J. (2014) Smart meter communication standards in Europe: a comparison. *Renewable and Sustainable Energy Reviews*, 43, 37.

33. Depuru, S.S.S.R., Wang, L., and Devabhaktuni, V. (2011) Smart meters for power grid: challenges, issues, advantages and status. *Renewable and Sustainable Energy Reviews*, 15, 2736–2742.

34. Deloitte (2015) *Analysis of smart meter system on European applications and ideas on Turkey applications.*

35. Aslam, W., Soban, M., Akhtar, F., and Zaffar, N.A. (2015) Smart meters for industrial energy conservation and efficiency optimization in Pakistan: scope, technology and applications. *Renewable and Sustainable Energy Reviews*, 44, 933–943.

36. U.S. Department of Energy (2014) Smart Grid System Report. 2014. doi: 10.1007/s13398-014-0173-7.2.

37. Khan, A.R., Mahmood, A., Safdar, A., Khan, Z.A., and Khan, N.A. (2016) Load forecasting, dynamic pricing and DSM in smart grid: a review. *Renewable and Sustainable Energy Reviews*, 54, 1311–1322.

38. Chakraborty, S., Ito, T., and Senjyu, T. (2014) Smart pricing scheme: a multi-layered scoring rule application. *Expert Systems with Applications*, 41, 3726–3735.

39. ComEd *Dynamic Pricing*. Available at https://hourlypricing.comed.com/live-prices/five-minute-prices/ (accessed June 15, 2017).

40. Kailas, A., Cecchi, V., and Mukherjee, A. (2013) A survey of contemporary technologies for smart home energy management. *Handbook on Green Information and Communication Systems*, Elsevier, p. 35–56.

41. Sanders, H., Kristov, L., and Rothleder, M.A. (2012) The smart grid vision and roadmap for California. *Smart Grid*, 127–159.

42. Ince, A.T., Elma, O., Selamogullari, U.S., and Vural, B. (2014) Data reliability and latency test for ZigBee-based smart home energy management systems. *7th International Ege Energy Symposium and Exhibition, Uşak, Turkey.*

43. Honda Smart Home. Available at http://www.hondasmarthome.com/ (accessed June 18, 2017).

44. Elma, O. and Selamogullari, U.S. (2015) A new home energy management algorithm with voltage control in a smart home environment. *Energy*, 91, 720–731.

CHAPTER 8

RESILIENT BATTERY MANAGEMENT FOR BUILDINGS

HOSSAM A. GABBAR[1,3] and AHMED M. OTHMAN[1,2]

[1]Faculty of Energy Systems and Nuclear Science, University of Ontario Institute of Technology, Oshawa, Canada
[2]Electrical Power and Machines Department, Faculty of Engineering, Zagazig University, Zagazig, Egypt
[3]Faculty of Engineering and Applied Science, University of Ontario Institute of Technology, Oshawa, Canada

8.1 INTRODUCTION

Battery management frameworks enable the users or the operators to control hardware battery systems when required for supplying various demands of the building. Upgraded controls improve the building's present frameworks and increment the capacity to oversee quality all through the building. A wide variety of frameworks and approaches have been proposed from researches to address the issue of increasing the utilization in private and business structures. These propositions depend on various correlative points of view, and frequently adopt an interdisciplinary strategy, which makes it difficult to get a far-reaching perspective of the cutting edge in the vitality administration of structures.

Energy management systems can spare vitality and enhance profitability by making an open workplace. Our reality is presently confronting two especially imperative patterns: rising derivative costs and worries about environmental change. Both make solid motivators for vitality protection. Sustainable development distinguished structures as one of primary vitality clients, which are expected to change vitality effectiveness. Structures represent 40% of essential vitality in many nations and utilization is rising.

Energy Conservation in Residential, Commercial, and Industrial Facilities, First Edition.
Edited by Hossam A. Gabbar.
© 2018 The Institute of Electrical and Electronics Engineers, Inc. Published 2018 by John Wiley & Sons, Inc.

Building energy management systems can spare vitality and enhance efficiency by making a satisfied workplace. Its stream enhances vitality of administration; in any case, normal building reviews and calibrating are important to guarantee the vitality administration is kept up.

This record condenses the specialized procedures for accomplishing vitality funds while enhancing inhabitant comfort. Energy systems advancement is subject to the physical plant, administrator, level of controls, and zoning, and in addition the sort of condition to which the framework is being connected. These data are focused to interior vitality investment funds usage experts, searching for an asset to direct them in evolving parameters, tuning building administration frameworks, and recommissioning existing frameworks.

8.2 EXPLORER OF SMART BUILDING ENERGY AUTOMATION (SBEA)

Smart Building Energy Automation (SBEA) is based on development of complete building energy management system linked to battery system in each apartment in building with resiliency algorithm, and to interface with the battery management.

SBEA has various features to achieve comprehensive framework:

- Building/home energy management system with user interface to manage and optimize performance of battery systems in apartments/buildings.
- Achieve battery performance and safety features.
- Achieve resiliency between apartments within one building and between two buildings, in a number of hazard scenarios.
- Integrate Smart Converters to integrate battery with PV/WT.
- Achieve IoT: Internet-of-Things for wireless connection with each system and monitor performance, and energy management with two-way control.

In addition, extra innovative features can be incorporated for more resiliency of the operation:

1. Incarcerating the battery lifetime, by operation limits incorporated with demand response to reduce the charging/discharging cycles.
2. Integrating the measurement platform with protection platform. Normally, the protective device work on the instantaneous values to trip or isolate, we can put predictive measuring online to send signals to the protective deceive to work on a preceded step; at this moment the control action is balancing load between battery or from utility electricity. And the last option is OCR action.

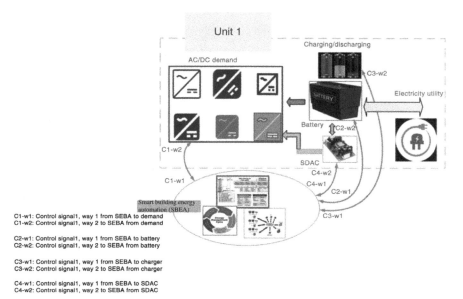

FIGURE 8.1   SBEA structure.

3. One important feature will be added to evaluate and asset the battery integration is measuring THD where the in/out/low/high operation scenarios lead to that, and active filter can be proposed to handle those and mitigate them (Figure 8.1).

8.3   SBEA SCOPES AND SPECIFICATIONS

A proposed SBEA with technical building management functions will have an impact on the energy performance of a building with the battery systems (Figure 8.2). Interfacing with Building Information Modeling (BIM) with virtual design to construct will have ability to size and placement components made easier. SBEA can achieve comprehensive functions as follows:

   (i) The detection of monomeric battery voltage
  (ii) The detection of battery temperature
 (iii) The detection of batteries operating electric current
 (iv) The detection of insulation resistance
  (v) The estimation of batteries SOC
 (vi) Communication with protection equipment
(vii) Communication with monitoring equipment

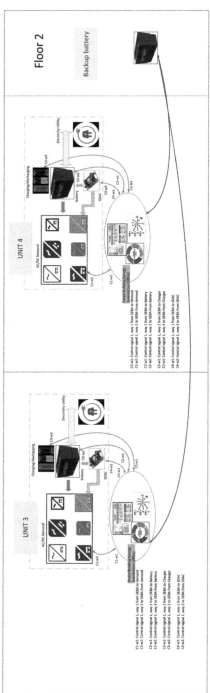

FIGURE 8.2 Integration of SBEA for building.

8.4 SBEA STRUCTURE

SBEA consists of central controller (SEBA00A-MC1), data collection modules (Volt/Tem module, current module, insulation resistance module), display, current sensor, and wires.

8.4.1 Connection Structure

Connection structure monitors total voltage, total current, remaining capacity (SOC), and highest temperature in a battery pack. It can display each cell voltage, a temperature collection point in a module. It also informs on how many cells a module manages, battery upper limit and lower limit beep warn voltage, battery upper limit and lower limit cutting off voltage, temperature upper limit beep warns, biggest recharging current, current upper limit beep warns, voltage difference beep warns, recharging times, SOH, SOC initialize, rated capacity, reserve capacity correction factor, and system clock.

8.4.2 Technical Specifications

• Power supply	DC24 V
• Range of voltage measuring	DC 0~+5 V
• Voltage measuring accuracy	±(0.3%RD + 0.2%FS)
• Voltage display resolution	1 mV
• Hall sensor	
• Current measuring range	0–500 A(1000 A)
• Current measuring accuracy	±0.5%
• Current display resolution	0.1 A
• Temperature measuring range	10–85
• Temperature measuring accuracy	±1
• Minimum sampling period (voltage)	0.5 s
• Minimum sampling period (current)	0.1 s
• A h accumulative total minimum period	0.1 s
• A h display accuracy	0.1 A h
• A h measuring upper limit	>1000 A h
• R measuring range	>2 MΩ
• R measuring accuracy	±10%
• The largest switch voltage	30 V_{DC}
• The largest switch current	1 A

8.5 SBEA CONTROL STRATEGY

The control strategy is numerically handled for various drive cycle data sets using dynamic programming (DP) (Figures 8.3–8.5). Trained using the DP results, an

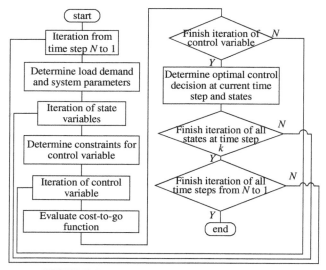

FIGURE 8.3   Flowchart of DP implementation.

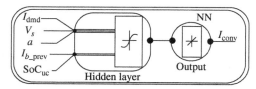

FIGURE 8.4   NN architecture for online energy management controller.

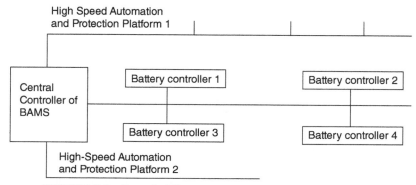

FIGURE 8.5   Smart building energy automation (SBEA) structure.

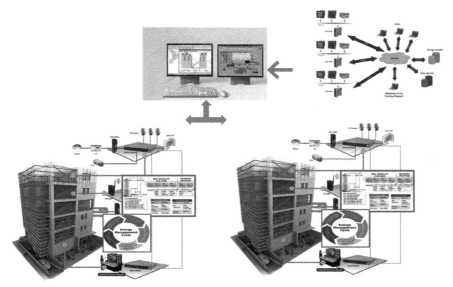

FIGURE 8.6   Integrated structure for building incorporated with SBEA.

effective and intelligent online implementation of the optimal power split, is realized based on neural networks (NNs).

8.6   COMMUNICATIONS AND DATA ANALYTICS

- Streams real-time data to external servers for analytics and trend data capture (Figure 8.6).
- Multisocket Ethernet interface allows concurrent operation of both local and remote operator panels, data analytics streaming, and Modbus TCP inverter control.
- Configurable bus interface for connection to inverters and chargers.
- Modbus TCP for connection to power conversion systems.

System_1 Parameter

Voltage alarm max limit	When recharging, one of cells voltage reaches the value; BAMS will warn and control charger to balance charging process
Voltage cut max limit	One of cells voltage reaches the value, will warn and stop charging. Only shut off the power, warning can stop
Voltage alarm min limit	When discharging, one of cells voltage reaches the value, BAMS will warn and controller to reduce out power. Avoid overdischarge. Voltage will be raised

Voltage cut min limit	One of cells voltage reaches the value, will warn and disconnect joint. Only shut off the power, warning can stop
Delta voltage alarm	The difference value between the highest voltage and the lowest voltage. It reaches the value, beep warns
Temperature alarm max limit	When the temperature is higher than the value, beep warns. It shares the joint with voltage upper limit warning

System_2 Parameter (Figures 8.7–8.9)

Max charge current	Output maximum current. The function is effective for the chargers that can communicate with our BAMS
Max discharge current	BAMS allows the maximum discharging current. If more, will warn
Rating capacity	Rated capacity of single cell. Consistent with the SOC100%
Capacity calibrate	Notice current sensor fix direction. Amend the capacity loss during charging and discharging
Cycle times	Times of charge and discharge
SOH	Battery pack health status
SOC initialization	That setting for after normal charging at the first time

8.7 TECHNICAL SPECIFICATIONS

This standard describes the size, characteristics, technical requirements, and matters of SBEA:

- Measure instrument.
- *Voltage meter request:* The accuracy of the measuring instrument is not less than 0.5. The resistance is not less than 10 K/V.

SBEA Modules

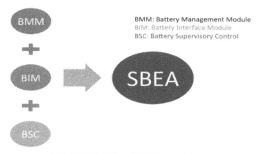

BMM: Battery Management Module
BIM: Battery Interface Module
BSC: Battery Supervisory Control

FIGURE 8.7   SBEA modules.

SBEA Architecture

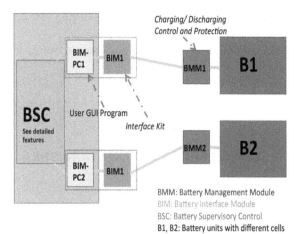

FIGURE 8.8 SBEA architecture.

SBEA Detailed Functions

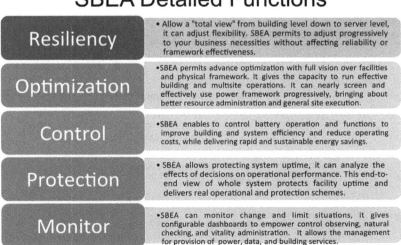

FIGURE 8.9 SBEA detailed functions.

- *Current meter request:* The accuracy of the measuring instrument is not less than 0.5.
- *Time meter request:* The accuracy of the measuring instrument is not less than 0.1%.
- *Temperature meter request:* The accuracy of the measuring instrument is not less than 0.5 °C.
- Temperature:15–35 °C.
- Relative humidity: 45–75% RH.
- Atmospheric pressure: 86~106 kPa.

8.8 SMART BUILDING ENERGY AUTOMATION: SBEA

8.8.1 Module Description

CDMB01 integrates DC charging function and protection circuit together on one board so that the user can build a Li-ion battery pack with full protection, charge battery pack by a conventional DC power supply or adaptor, and make Li-ion battery pack more compact and portable. The module function is to limit the charge current from any charge source (charger, charge controller, MPPT). If the charger source would attempt to charge the battery at a higher current, the charge current will be limited to the assigned amperes.

It will protect Li-ion battery pack from overcharging, overdischarging, and overdrain; therefore, it is must have to avoid Li-ion battery pack from exploding, firing, and damage. PCM with equilibrium function will have to keep each cell in better balance and good service life (Figures 8.10).

Charge Module Technical Specifications

No.		Item	Criterion
1	Voltage	Charging voltage	DC: 59.2 V
		Balance voltage for single cell	3.6V ± 0.03 V
2	Current	Balance current for single cell	58 ± 5 mA
		Current consumption for single cell	≤30 μA
		Maximal continuous charging current	30 A
		Maximal continuous discharging current	100 A
3	Overcharge protection	Overcharge detection voltage	3.9 V ± 0.025 V
		Overcharge detection delay time	0.7–1.3 S
		Overcharge release voltage	3.8 V ± 0.05 V
4	Overdischarge protection	Overdischarge detection voltage	2.0 V ± 0.062 V
		Overdischarge detection delay time	14–26 ms
		Overdischarge release voltage	2.3 ± 0.10 V

(*Continued*)

Charge Module Technical Specifications

No.	Item		Criterion
5	Overcurrent protection	Overcurrent detection current	120 ± 30 A
		Detection delay time	8–16 ms
		Release condition	Cut short circuit
6	Short protection	Detection condition	Exterior short circuit
		Detection delay time	230–500 µs
7	Resistance	Protection circuitry	≤ 30 mΩ
8	Temperature	Operating temperature range	−40 °C to +85 °C
		Storage temperature range	−40 °C to +125 °C

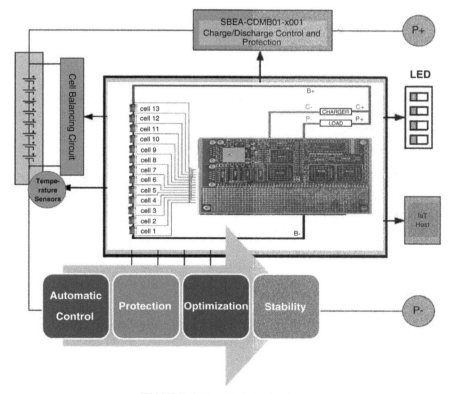

FIGURE 8.10   Module structure.

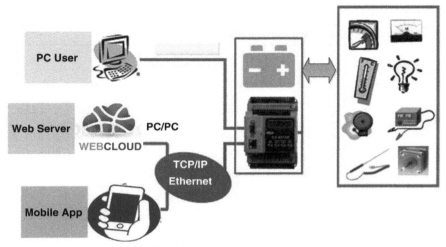

FIGURE 8.11 Module connections.

8.8.2 Standards

- IEC 61508, including E/E/PE system: IEC 61508 (Parts 1 °C 7): Functional of electrical/electronic/programmable related systems.
- IEC 61427-2: Secondary cells and batteries for energy storage – General requirements and methods of test.
- IEC 61850: Communication networks and systems in substations.
- IEEE Std. 1625 – 2008: Information technology equipment: General requirements.
- NF EN 50272: Safety requirements for secondary batteries and battery installations.
- NF EN 61982 – 2013: Secondary cells and batteries containing alkaline or other nonacid electrolytes – Safety requirements for portable sealed secondary cells, and for batteries made from them, for use in portable applications.
- NF EN 62133 – 2013: EEE Standard for Rechargeable Batteries for Multi-Cell Devices.
- U.S. National Electrical Code (NEC) 690.5 Compliant.
- Manufactured in a certified ISO 9001 facility (Figure 8.11).

8.9 SAVING WITH SOLAR AND BATTERY INTEGRATION

8.9.1 Residential Demands

Solar electric system saves grand resident over $1200 a year

Single family house: system size – 5.59 kW

Number of solar panels: 26 roof-mounted panels

Yearly energy production at 10,509 kWh

Net investment after incentives, rebates, and tax credits: $9385

Saving about: $1215 per year

Compared to net investment and this level of savings, payback will be less than 8 years from purchase date.

8.9.2 Commercial Demands

Store area $= 5000 \, \text{ft}^2$ store

Daily electricity $= 67 \, \text{kWh}$ average

Electricity monthly 2000 kWh

Average commercial electricity rate $=$ charges for use is 15 cents per kWh

Average spend $= \$300$ on the electric bill each month and about $3600 a year on electricity

Solar panels to produce 67,000 W daily

System size, solar panels that produce 250 W each per day

Number of solar panels: 268 solar panels

Total cost of approximately: $30,245

Saving: 45% of bill $= \$1620$

8.10 SBEA MAIN OBJECTIVES

SBEA manages and controls multiple-connected battery packs. It monitors the battery quantities SOC–SOH–DOD, monitoring and controlling key parameters of the operation for the following:

+ Protection of the cells/the battery packs
+ Maximize the battery lifetime
+ Match the battery with the demand requirements
+ Interface with the schemes of original power network

8.11 SBEA FUNCTIONS

- *Measurements:*
  - Cell voltages
  - Pack voltages
  - Pack current
  - Pack temperature
  - Pack state of charge estimation

- *Cell Internal Resistance:*
  - Important for calculating state of health
- *Pack Health and Maintenance:*
  - Cell balancing (active or passive)
  - Temperature management
- *Protection:*
  - Short-circuit protection
  - Overcurrent protection
  - Overvoltage limit protection
  - Undervoltage limit protection
  - Temperature cutoff protection
- *Data:*
  - Battery pack configuration
    - o Voltage threshold settings
    - o Current threshold settings
    - o Temperature threshold settings
  - Local storage (usually for error reporting)
  - External communication

8.12   CURRENT CONTROL MODULE: SBEA

- CCM is a module that is able to control the current entering the charge input of the battery.
- It does the following:
  - Regardless of available power, limit the charge current to a particular value (e.g., 30 A).
  - Have little to no impact on the voltage supplied to the system, and can stand charge voltages up to 62 V.
  - Have a high efficiency as not to burn or waste the available power (and ideally, not generate a lot of heat).
  - Be inserted into the charge line of the battery.

The discharge line of the battery would not be limited/impacted by this module.

8.13   PROTECTION PCM MODULES

- PCM is mounted separately from the cells of the battery.
- It provides the following functions:
  - Short-circuit protection

SBEA Schematic

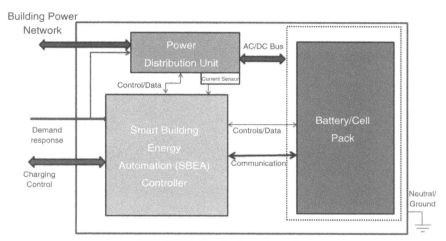

FIGURE 8.12   SBEA schematic.

- Cell balancing
- Overcurrent cutoff
- Undervoltage cutoff
- overvoltage cutoff
- When the battery is fully charged and the PCM disallows charge current, it will still permit discharge current. Likewise, when the battery is fully discharged and the PCM disallows further discharge current, it will still permit charge current.
- The protection remains as long as the PCM detects the voltage levels required to put it into a protective state, and releases when the PCM no longer detects the fully discharged or fully charged state (Figure 8.12).

8.14 MANAGEMENT CONTROL

- Cell voltage
- Cell balancing
- Cell temperature
- Fail-safe response to faulted conditions
- Sensor integrity and diagnostics

Battery/cell pack variables description includes the battery modules, voltage sensors, cell voltage balancers, temperature sensors, condensation sensors, and ground-fault detection circuitry (Figure 8.13).

Battery/ Cell Pack Variables

* Description:
  — Includes the battery modules, voltage sensors, cell voltage balancers, temperature sensors, condensation sensors, and ground fault detection circuitry

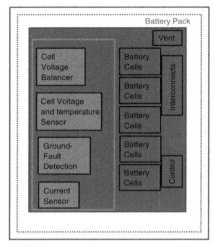

FIGURE 8.13 Battery pack schematic.

8.15 BATTERY MANAGEMENT AND CONTROL VARIABLES

* *Description:*
  - Calculations for SOC, SOH, DOD
  - Operation diagnostics and optimization

Battery Management and Control Variables

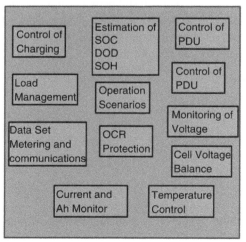

* Description:
  – Calculations for SOC, SOH, DOD
  – Operation Diagnostics and optimization
  – Data handle and Communication
  – Safety Control and Protection
* Function:
  – Manages the power direction from and to the battery pack based on the SOC
  – Manages on power exchange-related requests from other packs or sources

FIGURE 8.14 Control variables pack schematic.

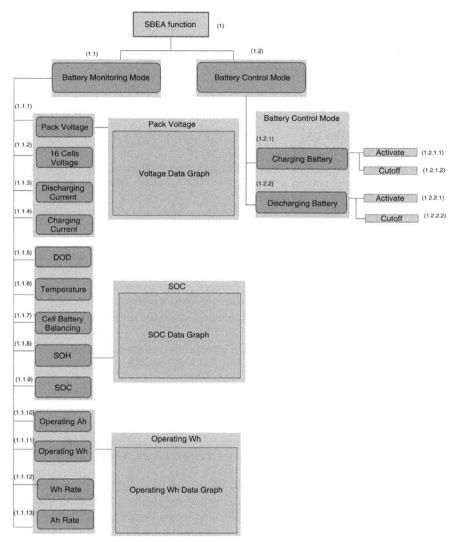

FIGURE 8.15 SBEA functions.

- Data handle and communication
- Safety control and protection
- *Function:*
  - Manages the power direction from and to the battery pack based on the SOC
  - Manages on power exchange-related requests from other packs or sources (Figure 8.14)

- *Benefits:*
  - End-to-end control solutions, reliable and scalable
  - Real-time operating system
  - Safe, secure, and reliable
  - Complements existing control and protection systems (Figure 8.15)

FURTHER READINGS

Cao, J. and Ali, E. (2012) A new battery/ultracapacitor hybrid energy storage system for electric, hybrid, and plug-in hybrid electric vehicles. *IEEE Transactions on Power Electronics*, 27 (1), 122–132.

Carrasco, J.M. et al. (2006) Power-electronic systems for the grid integration of renewable energy sources: a survey. *IEEE Transactions on Industrial Electronics*, 53 (4), 1002–1016.

Chen, H. et al. (2009) Progress in electrical energy storage system: a critical review. *Progress in Natural Science*, 19 (3), 291–312.

Daz-Gonzlez, F. et al. (2012) A review of energy storage technologies for wind power applications. *Renewable and Sustainable Energy Reviews*, 16 (4), 2154–2171.

Edwards, P.P. et al. (2008) Hydrogen and fuel cells: towards a sustainable energy future. *Energy Policy*, 36 (12), 4356–4362.

Emadi, A., Williamson, S.S., and Khaligh, A. (2006) Power electronics intensive solutions for advanced electric, hybrid electric, and fuel cell vehicular power systems. *IEEE Transactions on Power Electronics*, 21 (3), 567577.

Fuchs, G. et al. (2012) *Technology overview on electricity storage*. ISEA, Aachen.

Ibrahim, H., Adrian, I., and Jean, P. (2008) Energy storage systems characteristics and comparisons. *Renewable and Sustainable Energy Reviews*, 12 (5), 1221–1250.

Jensen, S. and Engebrecht, K. (2012) Measurements of electric performance and impedance of a 75 A h NMC lithium battery module. *Journal of the Electrochemical Society*, 159 (6), A791–A797.

Khaligh, A. and Zhihao, L. (2010) Battery, ultracapacitor, fuel cell, and hybrid energy storage systems for electric, hybrid electric, fuel cell, and plug-in hybrid electric vehicles: state of the art. *IEEE Transactions on Vehicular Technology*, 59 (6), 2806–2814.

Mclarnon, F.R. and Cairns, E.J. (1989) Energy storage. *Annual Energy Review*, 14, 24171.

Nature, F.L., Cher, M.T., and Michael, P. (2015) Effect of temperature on the aging rate of Li ion battery operating above room temperature. *Nature*, 2015, 12967.

Wiaux, J.P. and Chanson, C. (2013) *The lithium-ion battery service life parameters*. EVE IWG Session, Geneva.

CHAPTER 9

CONTROL ARCHITECTURE OF RESILIENT INTERCONNECTED MICROGRIDS (RIMGs) FOR RAILWAY INFRASTRUCTURES

HOSSAM A. GABBAR,[1,3] AHMED M. OTHMAN,[1,2] and KARTIKEY SINGH[1]

[1]Faculty of Energy Systems and Nuclear Science, University of Ontario Institute of Technology, Oshawa, Canada
[2]Electrical Power and Machines Department, Faculty of Engineering, Zagazig University, Zagazig, Egypt
[3]Faculty of Engineering and Applied Science, University of Ontario Institute of Technology, Oshawa, Canada

9.1 INTRODUCTION

Microgrid (MG) has brought remarkable change in the field of energy system. It has given the way to solve many problems faced by the traditional system and has also created some challenges to be met. One challenge is the control of MG and the other challenge created by MG is the control of MG. As in the case of control, microgrids also need a different protection scheme as traditional scheme gets useless mainly because of bidirectional power flow and power electronic interface. Grid-tied microgrids can supply its local load itself and can also take help of utility to meet this load. During parallel operation, control system is designed under which MG disconnects from the grid during various operating conditions and when the power quality of the grid is very low. In some cases, an energy storage unit is also provided in the microgrid to manage the control issues of the microgrid as well as to provide a better and reliable power source. Microgrids interoperate with existing power systems, information systems, and network infrastructure, and are capable of feeding power back to the larger grid during times of grid failure or

Energy Conservation in Residential, Commercial, and Industrial Facilities, First Edition.
Edited by Hossam A. Gabbar.
© 2018 The Institute of Electrical and Electronics Engineers, Inc. Published 2018 by John Wiley & Sons, Inc.

power outages. Integration of energy storage in microgrid has a significant impact on the power flow and the operating conditions of the utility equipment at the customer end. Depending on the type of energy storage and also on the grid connection type, these impacts could be positive to cause voltage quality criteria. Flywheel uses the mechanical form of energy, the kinetic energy of a fast spinning disk, which is the stored energy. Flywheels can be used to store energy for power systems when coupled to an electric machine. To retrieve the stored energy, the process is reversed – the motor acts as a generator powered by the braking of the rotating disc. Flywheels can be used in microgrid power quality application as it can charge and discharge frequently.

Distributed Generation (DG) units of MG contain either microsources producing DC type power or AC type power. DC type microsources are like PV and fuel cell and AC type microsources are like micro turbine, gas turbines, and IC engine-based generators. Both of these types of devices need power electronic interface to be connected to load [1,2]. There needs to be a control system for the safe and reliable operation of MG both in grid-connected mode and islanded mode. There are several ways of integrating these DGs to the grid.

In transportation electrification, electric power is supplied to the railway trains and buses without a local fuel supply. This system has many advantages but it needs a lot of financial investment. There are different criteria that lead to the selection of an electrification system including the economics of electrification type, maintenance requirements. Different systems are used depending upon the requirements [3,4]. Type of electrification in transportation can be classified mainly on the basis of the following factors: voltage, current, and contact system. On the basis of contact, this electric system can be categorized as overhead, third rail system, and fourth rail system. The modern railway system uses AC power from the grid and delivers it to the locomotive where it is used by DC or AC traction motors. If we use DC traction motor, then it is directly used otherwise in case of AC, it needs to be reconverted to AC. A traction electrification system's task is to provide electrical power to trains using power supply system, distribution system, and a return path for the power. AC and DC electrification systems are different in the sense that in DC system, each substation must have a transformer and a rectifier bridge because it supplies voltage in DC, so it needs to supply at a voltage that is suitable for the traction motors, which means that the distance between substations needs to be less; for high power requirements, current will be more and thus more losses and size of the conductor will also increase [5–8]. On the basis of these facts, it can be concluded that DC systems are more suitable for urban transportation with relatively short stops between the stations. The AC system is more suitable for the intercity operational purposes where distance between stops is large and higher speed is required. Electrical power is supplied from the utility to the substation and then transformed to voltage suitable for traction motors and then transmitted to the trains. In autotransformer fed system, utility power is transformed to voltage that is higher than the suitable voltage for the traction motors and this voltage is converted to suitable level along the route by autotransformers. Booster transformer is a modified directly fed system to reduce electromagnetic interference [9–12].

Many publications have been published to advance the operation and to administrate the resilient performance of the MGs. Likewise, they have discussed the ability for recovering the system stability from any operating conditions. Railway infrastructure is one of the main sectors that will take benefits from applying MG to spare and utilize the braking power with penetration of DGs with renewable energy and advanced energy storage systems. That strategy will prompt enhancement in the power quality and framework dependability, where as of now the traction power of the railway systems takes interest in providing the power. MG can be a solution with significant answer for all the operation scenarios as normal and emergency, cases with more constraint and concerns of quality and security. Reduction of power outage, increase of power reliability, and decrease of demand not survived (DNS) are objectives that can be achieved with MGs usage. In addition, the decrease in pollution and greenhouse gas emissions will be acknowledged [13–15]. In perspective of some research challenges, it is vital to assemble and show a working model for interconnected MG for transportation, particularly railroads with achieving the high level of standard for the operation and performance. Advanced classes of control frameworks should be presented to assess and affirm the specialized advantages. The proposed research will focus on integration of railroad infrastructure with MG application and coordinated renewable energy penetration in presence of recent energy storage systems and modern control frameworks. In that paper, the microgrid will be connected to deal with the power streaming between trains as a unit of railroad frameworks and the utility network, and furthermore, will deal with the operation of the energy storage units. The traded power from and to the utility grid will be influenced by the value duty of the power price. Along with the railroad framework, there will be microgrids that can accomplish those destinations with least cost. Microgrid energy streams are characterized as controlled plan, which will be mimicked on day-to-day availability.

9.2 PROBLEM STATEMENT

Resilient Interconnected Microgrid (RIMG) is presented with presence of advanced storage system and various Distributed Generations (DGs) along of the supplied utility grids. Controlled modeling platform for the whole system is presented for balancing the energy streams exchanged between train units while movement and brake power stream, storage power stream, and the primary power from utility grid as indicated in Figure 9.1a.

The objective formulation inside the scoped structure will rely upon getting the optimal power patterns from and to MG to supply the railroad and furthermore the capacity rate of the storage units that minimize the working expense in nearness of the MG. That ought to be accomplished by keeping the acceptable lines power and

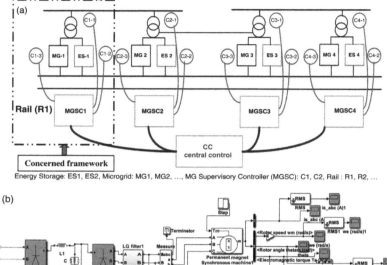

Energy Storage: ES1, ES2, Microgrid: MG1, MG2, ..., MG Supervisory Controller (MGSC): C1, C2, Rail : R1, R2, ...

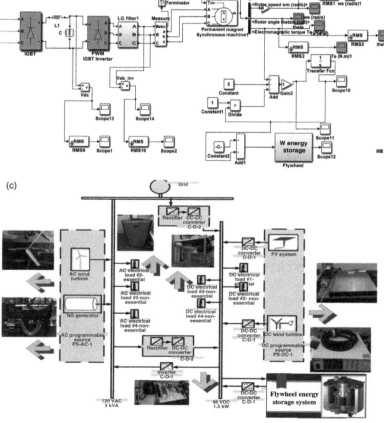

FIGURE 9.1 (a) General layout of RIMG for railway network. (b) Modeling blocks and structure of FESP. (c) Demonstration of MG in ESCL.

the voltage profile that will be considered as effective optimizer platform for the energy between the railway infrastructure and the microgrids.

The energy storage systems will be based on flywheel energy storage platform (FESP). FESP will have integrated schemes and technologies, as shown in Figure 9.1b. First, it will apply flywheel energy storage; flywheel storage systems are one of the most promising technologies as energy storage systems for a variety of applications, including automobiles, economical rural electrification systems, and stand-alone, remote power units. The integrated FESP can be applied as per optimality criteria. The flywheel system has advanced features as dynamic control scheme to enhance the dynamic performance and storage capability based on optimality criteria, both economic and technical considerations.

9.3 ESCL MG PROTOTYPE

Figure 9.1c indicates MG design with various DG units that are providing distinctive AC and DC sides. In light of the yield of the arranging stage, the AC sources are DFIG wind turbine generator and small-scale gas turbine generator. The DC sources are battery, fuel cell stack in view of hydrogen and PV clusters. For full use operation, booster-action converters, AC/DC, DC/AC, and DC/DC converters are associated with the microgrid. Different sorts of DC and AC load can be connected to, for example, resistive burdens, mechanized dc arrangement, engine stack, direct AC loads, nonlinear AC burdens, and three-phase mechanized burdens.

The plan of MG with equipment exhibits is introduced in Energy Safety and Control Laboratory (ESCL), University of Ontario Institute of Technology (UOIT). Since 2011, the individuals from ESCL are taking a shot at MEG with more than 5 projects and 20 publications; many software packages such as Simulink/SimPower with m-file codes and programming are applied with the hardware installations to perform and prove the digital excitations and simulations of the concerned MG network.

9.4 MICROGRID SUPERVISORY CONTROLLER

The methods proposed in this study are divided into several cases. As is well known, microgrid has two modes of operation, namely, islanded and grid-connected mode with different cases associated with the operation of a microgrid. The general architecture of the microgrid with the battery in grid-connected mode is shown in Figure 9.2a.

The architecture of the microgrid with the MGSC is shown in Figure 9.2b. In this case, MGSC manages the flow of power and determines which sources should

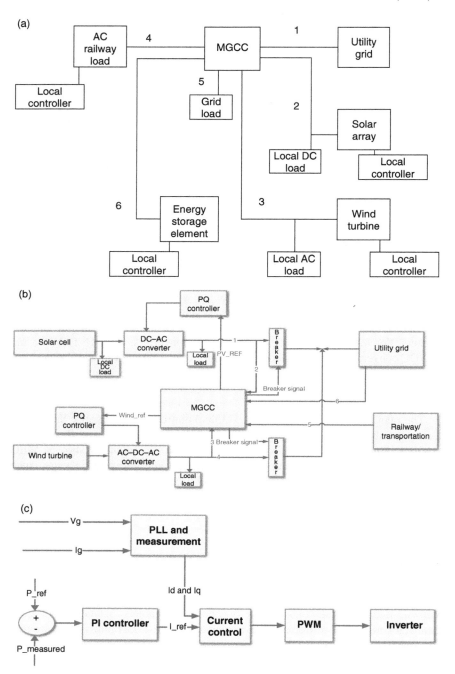

FIGURE 9.2   (a) General cases of the considered microgrid. (b) Microgrid structure with MGSC architecture. (c) PQ control strategy by current regulation. (d) Flowchart of MGSC strategy.

(d)

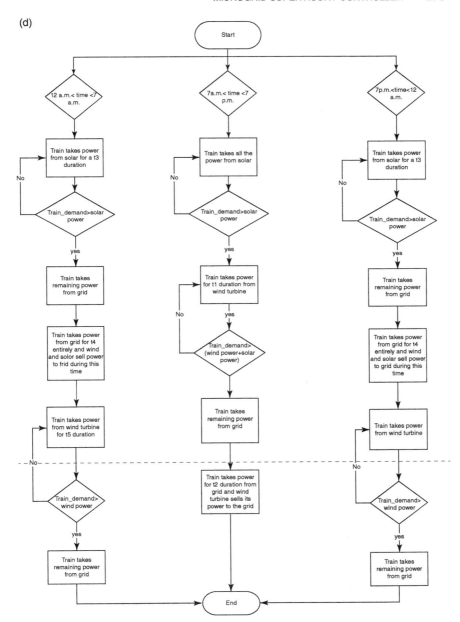

FIGURE 9.2 (*Continued*)

be used to meet the demand of transportation load and also determines the amount of power injected by the microsources to the grid. As shown in Figure 9.2b, there are two paths for power flow for microsources. One path is for power flow from the microsources to the grid or to the transportation load and is numbered as 2 and 3 in the figure and the second path is for when the microsources have no enough power to meet their local load and need power from the grid and is numbered as 1 and 4. For the subsequent operation of different paths, a three-phase breaker is provided in each path whose operation is governed by the MGSC. Also the power injected by the microsources is controlled by MGSC as MGSC provides the reference signal to PQ controllers attached to the converters. Signals are named as pv_ref and wind_ref. This being said, it's time to look into the architecture of the microgrid central controller. The architecture of the MGSC is shown in Figure 9.2b. The MGSC architecture shows that each microsource has further two paths: one for the power injection to the grid and the other for supplying power to the transportation load. Each path has a three-phase breaker for the subsequent operation. The breakers are governed by the control signals supplied by the controller as shown. Note that there is also a path between the grid and the transportation load governed by the controller.

Controller also takes the predicted load demand of each side as input and the declared rates of buying and selling the power from and to the grid and railways.

9.5 CONTROL STRATEGY

In the grid-connected mode, energy storage unit may or may not be considered depending upon the cost of energy storage element. Use of MGSC depends upon the business aspect of the system. Similarly, the design criteria of MGSC also depend upon the business aspect of the system. In this system, the economy of the system has been considered as the design criteria of the microgrid central controller.

The controller as shown in Figure 9.2 consists of phase-locked loop, active power control, and current control. The active power control loop uses a PI controller to minimize the error between the power injected and the reference power and in process generates the reference value of direct component of the current and note that the quadrature component of the current is set as zero, meaning zero reactive power. Now the current controller also uses a PI controller to bring the value of current measured through the phase-locked loop to the reference value of current generated by the power control loop and in this process, this generated a PWM signal, which is fed to the universal converter connected to the solar array. This control strategy is important because it makes sure through phase-locked loop and current control that all the power available at the terminal of the microsource gets injected to the grid. This control strategy imposes a problem if the DC voltage regulation is important at the inverter side as for different values of reference power, a different value of DC voltage will appear, which will make it unfit for the DC load, if any, before the inverter. To avoid this, a two-phase control strategy is implemented in the solar array side. Voltage control is implemented at the inverter side and the current control is implemented at the boost converter side, which effectively results as the power and voltage control strategy. The reference signal for PQ controller is generated by microgrid central controller whose operation is decided by the operator. In this chapter, the microgrid central controller has been taken in a fashion that minimizes the cost of the operation and maximizes the benefits to all the entities involved in the operation.

The following points will help understand the operation of MGSC:

- As stated earlier, the central controller makes decisions by taking inputs of loads, different rates of electricity during different periods of the day.
- The following flow chart explains how the central controller takes decisions.
- Since the rate of electricity differs for different times of the day, so the MGSC starts by checking which time is it during the day and thus gets the idea of the different rates during different times.

It compares between the rates and switches power between transportation load and different sources, that is, grid, wind turbine, solar cell to minimize the cost of entire operation and give maximum benefit to all system components.

9.6 SCENARIOS WITH SIMULATIONS AND RESULTS

The proposed scenarios are aimed at utilizing many software packages such as Simulink/SimPower with m-files codes and programming, which is tailored to the chosen MG analysis, and demanding profile. One-day profile for train traction utilization will be utilized as a part of the executed scenarios.

For this case, several different cases have been considered depending upon the travel time requirements. For example, a distance of 80 km can be travelled in either 30 or 90 min. In each case, depending upon the time, power requirement by the train will be different. Simulation time is taken as 5 s.

The results obtained will be discussed later. Active power demand of the train is shown in Figure 9.3a and b.

As can be seen, when train is drawing power, during part of the time power is supplied by grid, solar cell, and wind turbine. These intervals are decided by optimization algorithm. For that moment, a rough optimal value has been estimated. Figure 9.3c shows the power supplied from solar cell to the train.

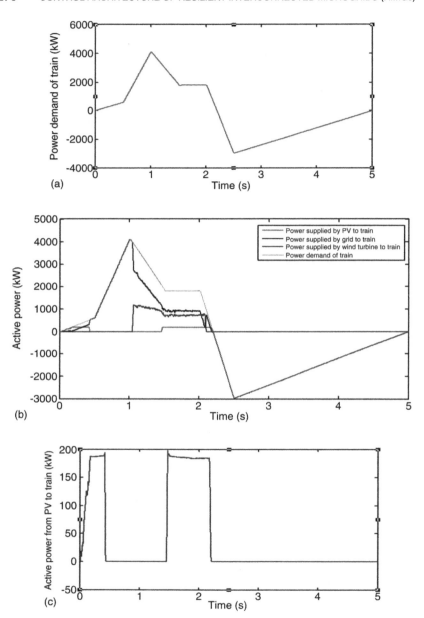

FIGURE 9.3 (a) Active power demand of the train. (b) Active Power supplied and consumed by train from different sources. (c) Active power supplied by solar array to the train. (d) Active power sold by solar array to the grid.

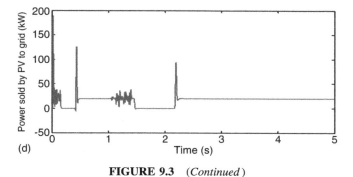

(d)

FIGURE 9.3   (*Continued*)

Figure 9.3d shows the power sold by solar cell to the grid. Note that reference power is set to 20 kW and the power injected is also very close to the reference power, where Figure 9.4a shows the total power supplied by solar cell.

Figure 9.4b shows the power supplied by the grid to solar panel, Figure 9.4c shows the reference value supplied by MGSC to solar cell PQ controller in per unit, and Figure 9.4d shows the power supplied by wind turbine to the train.

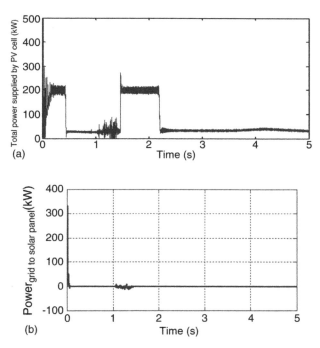

(a)

(b)

FIGURE 9.4   (a) Total active power generated by PV cell. (b) Power supplied by the grid to solar panel. (c) The reference value supplied by MGSC to solar cell PQ controller in per unit. (d) Power supplied by wind turbine to the train.

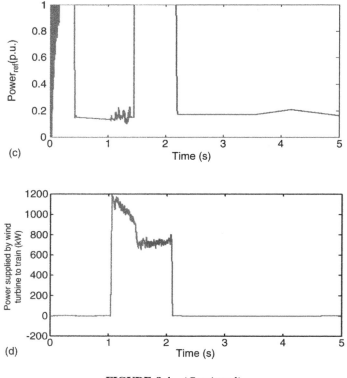

(c)

(d)

FIGURE 9.4   (*Continued*)

Figure 9.5a shows the power sold by wind turbine to the grid. Note that reference power is set as 500 kW. The difference between reference power and power sold to the grid in starting is due to the response time of controller, which is

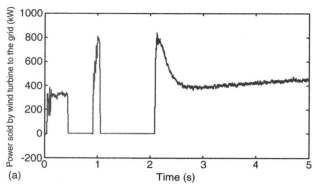

(a)

FIGURE 9.5   (a) Power sold by wind turbine to the grid. (b) Total power generated by wind turbine. (c) Power supplied by the grid to the wind turbine side.

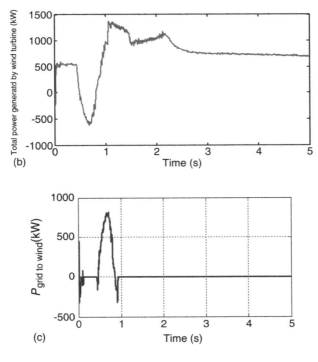

FIGURE 9.5 (*Continued*)

around 2 s. This response time, when on the actual scale of 24 h, is very decent but in here 24 h has been scaled to 5 s due to longer simulation periods.

Figure 9.5b shows the total power generated by wind turbine. Again note that response is in seconds, while the load profiles have been scaled from 24 h to 5 s. Figure 9.5c shows the power supplied by the grid to the wind turbine side.

Figure 9.6a shows the reference value supplied by MGSC to wind turbine PQ controller in per unit. Figure 9.6b shows the power exchange between the grid and the train.

9.7 COST AND BENEFITS

Installation cost of solar panels and wind turbines can be reviewed in Canada in Ref. [16,17]. A solar panel of capacity >50 kW costs roughly around 3 $/W. So a 200 kW solar panel will cost us roughly around $600,000; as far as wind turbines are considered, complete installation of wind turbine of capacity around 2 MW costs us around 3–4 million. So wind turbine that is considered here is 1.5 MW and will therefore cost roughly around $3 million. Associated costs of buying and

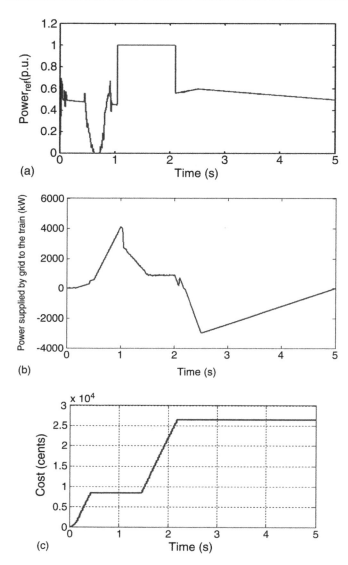

FIGURE 9.6 (a) The reference value supplied by MGCC to wind turbine PQ controller in PU. (b) Power exchange between the grid and the train. (c) Cost of power sold by solar panel to the train. (d) Cost of power sold by wind turbine to the train.

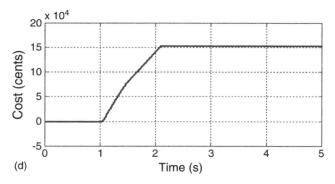

(d)

FIGURE 9.6 (*Continued*)

selling power to the transportation and grid were previously discussed. Assuming that microsources operate for 24 h, payback period of wind turbine and solar array can be calculated. The final costs are shown in Table 9.1. Payback period of wind turbine and solar panel is shown in Table 9.2.

Cost during the 24 h, which has been scaled to 5 s, for power exchange between different elements is shown in Figure 9.6c and Figure 9.6d. Figure 9.6c shows the cost of power sold by solar panel to the train. Figure 9.6d shows the cost of power sold by wind turbine to the train.

TABLE 9.1   Cost Involved in the Operation of Microgrid

Cost ($)	Power Exchange Direction
265.8	Solar panel to the train
101.9	Solar panel to the grid
367.7	Total cost earned by solar panel
1525	Wind turbine to the train
2305	Wind turbine to the grid
3830	Total cost earned by wind turbine
−690	Grid to the train
1100.8	Total cost paid by train

TABLE 9.2   Payback Period

Source	Payback Period (years)
Solar panel	4.47
Wind turbine	2.146

Figure 9.7a shows the cost of power sold by grid to the train. Negative cost indicates that train has compensated the cost from the grid and earned benefits. Figure 9.7b shows the total cost train has to pay.

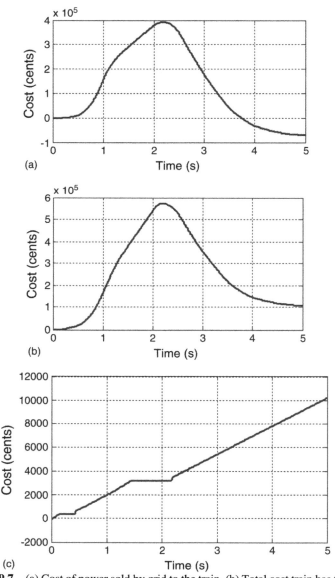

(a)

(b)

(c)

FIGURE 9.7 (a) Cost of power sold by grid to the train. (b) Total cost train has to pay. (c) Cost of power sold by the solar panel to the grid. (d) Total cost of power sold by solar panel.

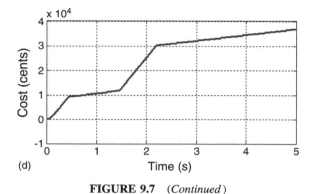

(d)

FIGURE 9.7   (*Continued*)

Figure 9.7c shows the cost of power sold by the solar panel to the grid. Figure 9.7d shows the total cost of power sold by solar panel.

Figure 9.8a shows the cost of power sold by the wind turbine to the grid. Figure 9.8b shows the cost of total power sold by wind turbine.

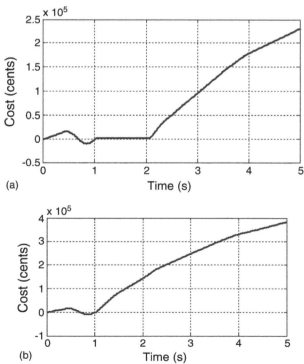

(a)

(b)

FIGURE 9.8   (a) Cost of power sold by the wind turbine to the grid. (b) Cost of total power sold by wind turbine.

9.8 CONCLUSIONS

In this chapter, Resilient Interconnected Microgrids (RIMGs) with hybrid Distributed Generation Resources (DGR) is proposed to be integrated with transportation infrastructure. RIMG will be incorporated with advanced energy storage units through the application of FESP. FESP will have integrated schemes and technologies to ensure optimum and highly efficient energy system. One contribution application is the joining of the transportation frameworks as another green-energy technology. Then, the chapter proposes Microgrid Supervisory Controller (MGSC), which will manage the flow of power and determine the sources involved to supply the demand of transportation load and also determine the power from RIMG to the utility network. MGSC will empower interconnected MGs to work straightforwardly with the advanced storage units. That the cost of power sold by grid to the train will be emphasized with negative cost indicates that train has compensated the cost from the grid and earned benefits. Digital simulation scenario has been validated by real data; the achieved results show the impact and effectiveness of the proposed strategies. The scenarios with simulation and results situation prove that the accomplished outcomes demonstrate the effect and viability of the proposed techniques.

REFERENCES

1. Lassete, R.H., *Microgrids*, College of Engineering, Wisconsin University, Madison, WI.
2. Lasseter, R.H. et al. Integration of Distributed Energy Resources: The CERTS Microgrid Concept. White paper. Available at http://certs.lbl.gov/pdf/50829-app.pdf.
3. Peng, Z.F., Yun W.L., and Leon M.T. (2009) Control and protection of power electronics interfaced distributed generation systems in a customer-driven microgrid. *IEEE Power and Energy Society General Meeting (PES'09)*, IEEE.
4. Katiraei, F. and Iravani, M.R. (2006) Power management strategies for a microgrid with multiple distributed generation unit. *IEEE Transactions on Power Systems*, 21 (4), 1821–1831.
5. Oyarzabal, J. et al. (2005) Agent based microgrid management system. *2005 International Conference on Future Power Systems*, IEEE.
6. Aris, L.D. and Hatziargyriou, N.D. (2005) Operation of a multi agent system for microgrid control. *IEEE Transactions on Power Systems*, 20 (3), 1447–1455.
7. Gaber, H., Othman, A.M., and Singh, K. (2016) *3rd International Conference on Power and Energy Systems Engineering (CPESE 2016)*, September 2016, Kitakyushu, Japan, pp. 352–359.
8. Soni, K.C. and Belim, F.F. (2016) MicroGrid during grid-connected mode and islanded mode: a review. *International Journal of Advanced Engineering and Research Development*, e-ISSN: 2348- 4470.
9. Stefano, B. et al. (2002) Control techniques of dispersed generators to improve the continuity of electricity supply. *IEEE Power Engineering Society Winter Meeting*, vol. 2, IEEE.

10. Pecas Lopes, J.A., Moreira, C.L., and Madureira, A.G. (2006) Defining control strategies for microgrids islanded operation. *IEEE Transactions on Power Systems*, 21 (2), 916–924.

11. Alfred, E. (2005) Applicability of droops in low voltage grids. *International Journal of Distributed Energy Resources*, 1 (1), 1–6.

12. Nikkhajoei, H. and Lasseter, R.H., *Microgrid Protection*, IEEE.

13. Buigues, G. Dyśko, A., Valverde, V., Zamora, I., and Fernández, E. (2013) Microgrid protection: technical challenges and existing techniques. *International Conference on Renewable Energies and Power Quality (ICREPQ'13)*.

14. https://en.wikipedia.org/wiki/Railway_electrification_system.

15. https://en.wikipedia.org/wiki/List_of_current_systems_for_electric_rail_traction.

16. Abramov, E.Y. Schurov, N.I., and Rozhkova, M.V. (2016) Electric transport traction power system with distributed energy resources. *Material Science and Engineering*, 127, 012001.

17. http://www.windustry.org/how_much_do_wind_turbines_cost.

CHAPTER 10

NOVEL LIFETIME EXTENSION TECHNOLOGY FOR CYBER-PHYSICAL SYSTEMS USING SDN AND NFV

JUN WU and SHIBO LUO

School of Electronic Information and Electrical Engineering, Shanghai Jiao Tong University, Shanghai, China

10.1 INTRODUCTION

Cyber-physical systems (CPS) refer to a system in which information processing and physical dynamics are closely integrated, which is an important component that pave the path to smart and efficient industry ecosystem. It brings sensing, computation, and communication to electrical systems with intelligence and capabilities. CPS has been widely used in many areas, including medical devices and systems, transportation vehicles, defense systems, factory automation, and so on. CPS usually consists of lots of sensors (mostly wireless sensors) and some controllers. By working together, sensors cooperate with each other to finish a common task. The collaboration of sensors is the critical requirement for CPS in industrial applications [1,2]. While parts of sensors in CPS lose their functions, the functions of industrial applications based on CPS are breached even if most sensors are still in work. The concept of collaboration lifetime of CPS is proposed to define the cooperative working time of the sensors in the same CPS network. As a result, extending the collaboration lifetime plays an important role in the CPS.

The inherent heterogeneity nature of CPS makes against the collaboration of sensors in CPS [2]. From the energy perspective, the sensors in CPS have different energy supply types, capabilities, and tasks. Some sensors perhaps can be powered by batteries and some others by, for instance, reused energy. This is one of the key

Energy Conservation in Residential, Commercial, and Industrial Facilities, First Edition.
Edited by Hossam A. Gabbar.
© 2018 The Institute of Electrical and Electronics Engineers, Inc. Published 2018 by John Wiley & Sons, Inc.

challenges to the collaboration of sensors in CPS. If the battery-based sensors are wearing out and they are not replaceable, the collaboration is breached. A potential way for extending the collaboration lifetime of CPS is to schedule the sensors' tasks [3,4]. A novel energy-oriented mechanism for extending collaboration lifetime of CPS is needed.

As one of the most important applications of CPS, wireless sensor networks (WSNs) have been studied for several years to improve their energy efficiency [5–9].Topology control is most mainstream and important technique used in WSNs to improve energy efficiency [5–7]. Topology control aims to control the graph representing communication links between nodes to reduce energy consumption. Sleep-mode-based energy efficiency technique is another important method [8,9]. By scheduling some unnecessary sensors into sleep mode, this technique will promote energy efficiency of WSNs.

At the same time, software-defined networking (SDN) and network function virtualization (NFV) are emerging as future network techniques that make the underlying networks and devices programmable [10–14]. SDN helps achieving global view of the network and controls the flows. By using virtual network function (VNF), the tasks of sensors can travel in the CPS. They help promote energy efficiency in CPS, where SDN is feasible to control topology and NFV helps CPS to implement the sleep mode, and then prolong the collaboration lifetime of CPS. Besides these features, the integration of SDN and sensor has been a trend [15–17]. And CPS has an excellent match with SDN because both of them have the controller component(s). All these invoke a method utilizing SDN and NFV to extend the collaboration lifetime of CPS.

On the other hand, various algorithms have been proposed to improve energy efficiency in WSNs. In Ref. [6] the authors present a clustering algorithm named low-energy adaptive clustering hierarchy (LEACH) to aggregate the data from sensors. The purpose of LEACH is to select sensors as cluster heads, so that the high energy dissipation in communicating with the base station is spread to all sensors in the sensor network. In Ref. [7] the authors propose a new distributed algorithm named scalable energy-efficient clustering hierarchy (SEECH), which selects cluster heads and relays separately based on sensors eligibilities. Simulation results demonstrate that all these works are effective in prolonging the network lifetime and supporting scalable data aggregation.

All these schemes have been proved effective in extending network lifetime. However, these algorithms only consider the topology control and do not take the nodes' task allocation into consideration. The topology decision problem based on the combination of SDN and NFV is not trivial, and these traditional methods cannot be applied directly. To tackle such complicated decision problems, game theoretical approaches are proposed [18,19].

This chapter proposes a novel lifetime extension scheme, NLES, for CPS by integrating SDN, NFV, and game theory seamlessly. The rest of the chapter is organized as follows: In Section 10.2, background and preliminaries are illustrated. The proposed mechanism is described in Section 10.3. Section 10.4 presents the game theoretic topology decision approach. Section 10.5 presents a case study. This chapter is concluded in Section 10.6.

10.2 BACKGROUND AND PRELIMINARIES

10.2.1 Topology Control and Sleep-Mode Techniques

There are several technologies for improving energy efficiency for wireless access communications. Among them, topology control is a sparkling method [5]. The idea of topology control is "to grant sensor nodes a sense of control over certain parameters such that these parameters can be manipulated in a way that benefit the network" [5]. Sleep-mode techniques selectively turn off some nodes in the network to reduce energy consumption according to policies. As an advantage, sleep-mode techniques are easier to test and implement because they are based on the current architecture and no replacement of hardware is required to use these techniques [8,9,12].

10.2.2 Game Theory

Game theory is "the study of mathematical models of conflict and cooperation between intelligent rational decision-makers" [18,19]. It is mainly used in economics, political science, and logic, as well as computer science. It was initially developed in economics that helped people understand large collections of economic behaviors. And now, game theory has played an important role in computer science. It is used to model interactive computations. Also, game theory offers a theoretical foundation to the fields of distributed systems such as WSNs and CPS. Specially, game theoretic approaches have been proposed to solve energy efficiency problem in WSNs [20,21]. In the applications, all the sensors play a game in the network simultaneously by picking their individual strategies. Typically, the WSNs are supposed to play an incomplete information game because the distributed nature of WSNs does not allow the sensors to have information about the strategies of other sensors.

10.3 PROPOSED MECHANISM

10.3.1 Assumptions

10.3.1.1 *Network Model Assumption*   Several assumptions of network model are made for the proposed mechanism.

- The CPS network is not restricted to static type. Namely, the components in CPS such as sensors, actuators, and controllers can move in the restricted area after deployment.

- Some sensors' functions can be replaced by other sensors. When multiple functions are aggregated into one sensor, all the transmitted data are compressed.
- Sensors can have access to low power sleep mode. Also, the transmission power can be adjusted.

10.3.1.2 Radio Model Assumption According to the previous research [6,7], the working current of radio is 28.8 times as much as that of microcontroller for a wireless sensor. This is to say that most of the energy is consumed by radio. Hence, power management is introduced to optimally schedule the active/sleep states of radio to avoid energy wastes. So, we focus on radio mode and communications between sensors using radio to provide energy-efficient mechanism.

This chapter uses the same radio model as illustrated in Ref. [6] for communication energy consumption, which is widely used in WSN research. It is beneficial to compare NLES with other famous energy-efficient schemes.

The required energy for transmitting a l-bit packet over distance d is

$$E_T(l,\ d) = E_{tx} \times l + E_{amp}(d) \times l \tag{10.1}$$

The parameter E_{tx} is the per bit energy consumption for transmission. $E_{amp}(d)$ is the energy required by the transmit amplifier to maintain an acceptable signal-to-noise ratio. $E_{amp}(d)$ is calculated using free-space model and two-ray model as follows:

$$E_{amp}(d) = \begin{cases} \epsilon_{fs} \times d^4, & d \le d_0 \\ \epsilon_{tr} \times d^2, & d > d_0 \end{cases} \tag{10.2}$$

where ϵ_{fs} and ϵ_{tr} denote transmit amplifier parameters corresponding to the free-space and the two-ray models, respectively. And d_0 is the threshold distance defined as follows:

$$d_0 = \sqrt{\epsilon_{fs}/\epsilon_{tr}} \tag{10.3}$$

The required energy for receiving a l-bit packet over distance d is

$$E_R(l,\ d) = E_{rx} \times l \tag{10.4}$$

The parameter E_{rx} is the per bit energy consumption for reception. Also, it is assumed that a data aggregation node spends E_{da} (nJ/bit per signal) amount of energy for its data aggregation.

10.3.2 Methodology for NLES

In this section, the methodology of this chapter is discussed. It includes the objective of NLES, subproblems derived from the objective, and the solutions for these subproblems. The basic idea of NLES is illustrated in Figure 10.1.

NLES provides a novel energy-oriented mechanism to extend collaboration lifetime of CPS based on topology control and sleep-mode techniques. To achieve this objective, there are two subproblems need to be solved. The first one is to provide a mechanism to support the implementation of NLES. The second one is to provide an approach to address the complicated decision problem in the mechanism effectively. And these two subproblems are not independent of each other. The mechanism is the basis for the approach. If the mechanism varies, the approach also needs to be modified.

SDN/NFV has inherent nature to improve the efficiency of topology control and sleep-mode algorithm. From the analysis of the energy consumptions, it is obvious that a more optimized topology and sleep mode, including data aggregation and compression, are applied using SDN and NFV. This scheme will consume less energy of the network than traditional methods. Also, the architecture of CPS has an excellent match with that of SDN, where both of them have controller role in their architecture. In this chapter, SDN and NFV are used to support the implementation of topology control and sleep mode for energy efficiency. SDN/NFV-based mechanism should be provided. A comprehensive mechanism includes the framework, details of the components in the framework, workflows, and protocols. It must be pointed out that no actual SDN equipment is introduced. The concept, framework, and protocols of SDN are adopted to utilize the advantages of SDN.

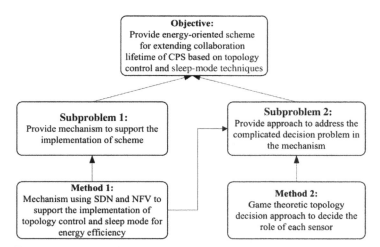

FIGURE 10.1 Basic idea of NLES. It presents the objective, subproblems, and methods.

There are mainly two types of decision parameters of the mechanism need to be calculated at run-time: topology clustering structure and VNF deployment plan. To simplify the description, topology clustering structure and VNF deployment plan are collectively called the "topology decision." The topology decision problem is not trivial, because it must consider topology control and task allocation simultaneously. Different from traditional WSNs, sensors in CPS could have the complete information of other sensors, because CPS has controller role and sensors must communicate with the controller. To obtain an effective topology decision, the approach must utilize the advantages of SDN and NFV fully and tackle the difficulties in complicated decision. A game theoretic approach is used to make decision of the mechanism, because it has proved effective in making complicated decision. Controller acts as a decision maker and acquires prerequisite information from sensors. It contributes to making more precise and effective topology decision.

10.3.3 The Proposed Framework

In this section, a new architecture that utilizes SDN and NFV to improve energy efficiency for CPS is illustrated. The new architecture includes a logical controller, enhanced sensors, and actuators besides the traditional components in CPS. The new architecture is shown in Figure 10.2.

In the enhanced architecture, there is a logical central controller. The logical central controller is based on the controllers in CPS, which not only control the network topology but also control the modes of sensors besides the common

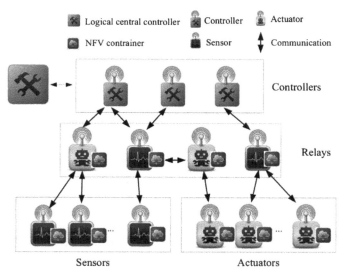

FIGURE 10.2 The proposed CPS network architecture. The architecture includes controllers, relays, sensors, and actuators.

functions of traditional CPS controller. The logical central controller is implemented as several physical controllers.

All the sensors and actuators run embedded operation systems and virtual machines that are called VNF containers. Based on embedded operation systems and virtual machines, the sensors and actuators are able to install VNF. It is practical because of the improvement of the capabilities of the processors and storages in current sensors. And the key challenges in the implementation of NFV are overcome, as illustrated in Ref. [22]. More details about the controller and sensors are illustrated in Figure 10.3.

Besides the common functions of SDN and CPS controllers, the controller in NLES also has information gathering, topology decision, flow table, and virtual network functions (VNF). VNF could be realized by interpreted language scripts such as Java. Sensors in NLES have four classic components: power unit, sensoring unit, processing unit, and transceiver unit [13]. With the current powerful processors and storages of sensors, VNF containers and flow table are realized in the proposed architecture that is different from common sensors. Additionally, the transmission power is controlled by a radio controller in each sensor.

The VNF registry functional entity maintains all the information about the VNFs that have been registered. The information includes function, required

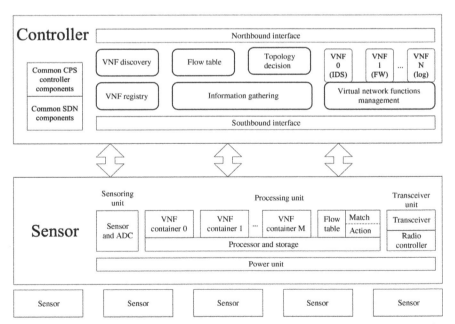

FIGURE 10.3 Details of controllers and sensors in the proposed architecture. The controller is the integrated version of SDN controller and CPS controller. Sensors have embedded operation systems and virtual machines.

resource, cost, effect, and so on. The VNF discovery functional entity provides VNF selection and negotiation capabilities. It interacts with the VNF registry functional entity to select the VNFs that match security criteria. The VNF management functional entity collects the security events from the VNF nodes in the domain. Also, it notifies the VFN nodes to perform the security policy.

After the controller makes topology decision using the proposed approach, the sensors enforce the decision utilizing the architecture. Besides the traditional functions of a classic sensor, the sensors have the traffic forwarding function in the proposed architecture. When all the sensors enforce the decision, they form the backbone network using their radios. They act as the switches in the CPS network.

When a sensor is selected as a sleeping sensor, it stops all the VNF containers and only keeps an idle thread alive. If the selected sensor has not installed the necessary VNF, it can retrieve the VNF source code from the controller and run it in its VNF container. It makes the processor consume low energy. In addition, it adjusts the radio to sleep mode to save energy. And the controller installs the VNF instances of the sleeping sensor into the peer of its replaceable pair or notifies the peer to perform specific VNF installed in the peer. Also, additional data aggregation VNF instances are installed into the peer or directly performed.

When a sensor is selected as a cluster head, it adjusts its transmission power to cover all of its cluster members and table flows are installed. Similarly, when a sensor is selected as a relay node, it adjusts its transmission power to cover its entire cluster heads and installs table flows. Other sensors are installed with its corresponding VNF instances and flow tables. And their transmission power is adjusted to reach their cluster head respectively.

10.3.4   Workflow at Run-Time of the Proposed Mechanism

The workflow at run-time of the proposed mechanism is illustrated in Figure 10.4. The workflow is proposed as a loop. After start-up, all sensors in the proposed

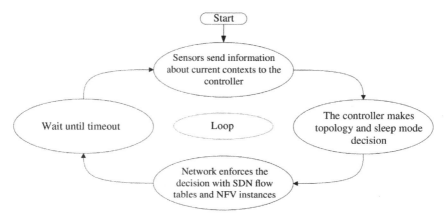

FIGURE 10.4   Workflow at run-time of the proposed mechanism.

mechanism send their information to the controller in each round, including residual energy, neighbors, location, instance functions, and so on. Utilizing these information, the controller makes topology decision based on game theoretic approach. After the decision is made, the network enforces the decision with SDN flow tables and NFV instances deployment. The workflow is triggered when next timeout occurs.

10.3.5 Messages Exchange Protocol between the Controller and Sensors

In the proposed mechanism, messages are exchanged between the controller and the sensors in CPS to negotiate the parameters that be used by the controller and the sensors as discussed in workflow. The protocol of exchanging messages is illustrated in Figure 10.5.

At the beginning of each round, the controller sends NOTIFY message to ask sensors to send their information to the controller. The sensors send INFORMA-TION messages, including information about themselves and their contexts to the controller. When all the information is collected, the controller makes decision about topology and roles of sensors. And then, the controller sends VNF_LABEL messages to the corresponding sensors to distribute the labels of VNF. If the VNF is already residing in the sensor, the sensor will send a CONFIRM message to the controller. Otherwise, the sensor will send a RETRIVE message to the controller to ask the controller to send VNF instances to it. And the controller sends the VNF

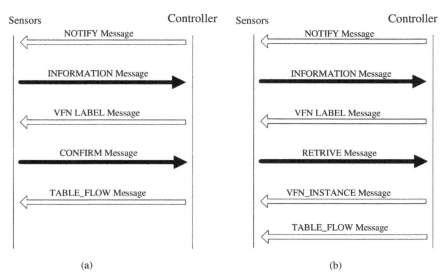

FIGURE 10.5   (a) Messages exchanged between the controller and VNF sensors. (b) Messages exchanged between the controller and non-VNF sensors.

instances to the sensor by VNF_INSTANCE message. At last, the controller sends TABLE_FLOW message to the sensor to control the topology.

10.4 GAME THEORETIC TOPOLOGY DECISION APPROACH

By using SDN and NFV, the proposed mechanism has advanced fundamentals to utilize the advantages of the combination of topology control and sleep mode. Consequently, complying with the advantages of SDN and NFV, topology decision method is necessary to be provided to complete the solution. Topology decision problem is formulated as a game and a survey of these works is presented in Ref. [20].

10.4.1 Problem Formulation

A game of topology decision is an interactive decision-making process between the sensors in CPS.

Definition 10.1

Topology decision game. A topology decision game is defined as a tuple $\mathbf{G} = (\mathbf{P}, \mathbf{A}, \mathbf{U})$. In the definition, $\mathbf{P}$ denotes players (namely sensors), $\mathbf{A}$ denotes their actions, and $\mathbf{U}$ denotes their utility functions.

10.4.1.1 Players In the topology decision game, the players are the total sensors in the network defined as follows:

$$\mathbf{P} = \{s_i, \quad \forall i \in (1, 2, \ldots, N)\} \tag{10.5}$$

where s_i represents the ith sensor and N is the total number of sensors in the network.

10.4.1.2 Actions The actions of sensor s_i can be split into two subtypes. The first subtype is the role space R_i of the ith sensor defined as follows:

$$R_i = \{r_n, \quad \forall n \in (1, 2, 3)\} \tag{10.6}$$

In NLES, there are three roles for sensors at run-time: cluster head, relay node, and common node. The sensors may act as one of these three roles at run-time.

And the second subtype is the executable task space T_i of the ith sensor defined as follows:

$$T_i = \{t_j, \quad \forall j \in (1, 2, \ldots, N_j)\} \tag{10.7}$$

where t_i represents the jth executable task of the ith sensor and N_j is the total number of executable task of the ith sensor.

When a sensor selects an action, it must combine these two subtypes of actions. So, the actions A_i of sensor s_i can be defined as follows:

$$A_i = \{a_m, \quad \forall m \in (1, 2, \ldots, N_m)\} = R_i \times T_i \tag{10.8}$$

10.4.1.3 *Strategy Profile* The game strategy profile Ψ of $\mathbf{G} = (\mathbf{P}, \mathbf{A}, \mathbf{U})$ is derived from the sensors' actions that is described as follows:

$$\Psi = A_1 \times A_2 \times \cdots \times A_N \tag{10.9}$$

10.4.1.4 *Utility Functions* The utility function of each sensor is the preservation of the sensor's lifetime when it performs a certain action defined as follows:

$$U_i = f(a_m), \quad \forall m \in (1, 2, \ldots, N_m) \tag{10.10}$$

The function of $f(a_m)$ should be chosen according to real applications.

10.4.1.5 *CPS Utility* U_i reflects the preservation of the sensor's lifetime. The utility of CPS is formulated as follows:

$$U_{\mathrm{CPS}} = \min_{1 \leq i \leq N} U_i \tag{10.11}$$

The equation indicates that the collaboration lifetime of CPS is decided by the minimal preservation of the sensors' lifetime. The bigger the minimal preservation, the longer the collaboration lifetime of CPS. Each sensor needs to maximize its lifetime and will attempt to select an action schedule that maximizes its preservation of lifetime as the best strategy in response to the topology decision. To extend collaboration lifetime of CPS, the sensors will attempt to schedule their actions not only to maximize the lifetime itself but also to maximize the collaboration lifetime of CPS, which means to maximize the CPS utility as defined in Equation 10.11. This is a noncooperative game [18,19].

10.4.2 Existence of NE

After the strategy profile, CPS utility, and the like are defined, a game is played by all the sensors simultaneously to decide their individual strategies. The combination of the individual strategies results in some strategy profile in Equation 10.5. Consequently, they select their actions to maximize their utilities. It is to say NE is achieved, when there is a set of actions with the property that no sensor can benefit when it changes its action solely while the other sensors keep their actions unchanged. NE is constituted of the set of actions and the corresponding utilities. NE is the solution of topology decision game [18,19]. But NE does not always exist. It exists if and only if a fixed point of a particular best response correspondence exists. To get topology decision, it is necessary to first verify the existence of NE in the proposed approach. The best response correspondence of the ith sensor is defined as follows.

Definition 10.2

Action $a_i^* \in \text{BR}(a_{-i})$ if

$$U_i(a_i^*, a_{-i}) \geq U_i(a_i, a_{-i}); \quad \forall a_i \in A_i \tag{10.12}$$

where a_{-i} denotes the actions selected by other sensors excluding the ith sensor defined as follows:

$$a_{-i} = \{a_1, \ldots, a_{i-1}, a_{i+1}, \ldots, a_N\} \tag{10.13}$$

Definition 10.3

$\bar{a}$ is a NE action profile if

$$a_i \in \text{BR}(a_{-i}); \quad \forall i \in (1, 2, \ldots, N) \tag{10.14}$$

where $\bar{a} = \{a_1, \ldots, a_N\}$.

Definition 10.3 shows that no sensor wants to change its action if other sensors do not change their actions. But this stable solution does not guarantee that the utility is optimal. According to the conclusion in Ref. [18], $\mathbf{G} = (\mathbf{P}, \mathbf{A}, \mathbf{U})$ is a potential game. Based on the properties of potential games and NE, $\mathbf{G} = (\mathbf{P}, \mathbf{A}, \mathbf{U})$ will converge to a conscious between sensors, which maximize the utility function. So, NE exists in the proposed approach.

10.4.3 Game Procedure

Based on the conclusion of the previous section, there exists NE in the proposed game. The game procedure of the proposed approach is presented in Algorithm 10.1.

Algorithm 10.1. Game procedure

Initialization: Each sensor creates action schedule as initial strategy and exchange it with other sensors.

On detection of timeout, execute:

For $i = 1$: N

The ith sensor gets optimization and obtains the utility and schedule strategy.

If the optimized schedule strategy is different from the previous schedule strategy

Send the new strategy information to other sensors.

else

Remain silent.

End If

End For

Repeat **For** until no more improvement for all sensors.

10.5 EVALUATION AND ANALYSIS

10.5.1 Algorithms Evaluation Setup

The evaluation is based on the result in Ref. [7]. To evaluate the effect of the proposed mechanism, two well-known energy efficiency methods are adopted to be compared with the proposed mechanism. These two methods are LEACH and SEECH. The parameters of simulation setup are listed in Table 10.1, which were used by LEACH and SEECH. It is beneficial to compare the effects. Three scenarios with different number of replaceable sensors and data aggregation compression ratio are introduced. The aim of these scenarios is to find the impact of these two parameters for energy efficiency in the proposed mechanism. The three scenarios are listed in Table 10.2.

In Table 10.2, the number of replaceable sensors indicates the amount of the sensors whose tasks can be replaced of by other sensors. D_{com} is defined to denote the data compression ratio in data aggregation for replaceable sensor nodes as follows:

$$D_{com} = \frac{D_i + D_j}{D_{total}} \tag{10.15}$$

TABLE 10.1 Parameters of Simulation Setup

Parameter	Value	Parameter	Value
Area	(0, 0)–(200, 200)	Initial energy	1 J
Location of data sink	(100, 350)	E_{tx}	50 nJ/bit
N	1000	E_{rx}	50 nJ/bit
K_{CH}	3	ϵ_{fs}	10 pJ/(bit m^2)
K_{CHC}	8	ϵ_{tr}	0.0013 pJ/(bit m^4)
K_R	10	d_0	87 m
K_{RC}	15	E_{da}	5 nJ/bit per signal
RNG	55	Packet size	4000 bits

TABLE 10.2 Parameters of Three Scenarios

Scenario	Number of Replaceable Sensors	Data Aggregation–Compression Ratio
1	400	1
2	500	1.11
3	500	1.25
4	200	1.01

where D_{total} denotes the total amount of the output data after data aggregation, D_i denotes the amount of the output data of the ith node before data aggregation, and D_j denotes the amount of the output data of the jth node before data aggregation.

10.5.2 Algorithms Evaluation Results

The numbers of "alive sensors" during simulation time in terms of round is used to evaluate the effects of the algorithms. An "alive sensor" is a sensor whose battery is not completely depleted. The numbers of "alive sensors" are derived by average results obtained from 200 simulation test. In each simulation test, a new distributed set of sensors are deployed.

Figure 10.6 illustrates the results. From Figure 10.6, the conclusion can be drawn that the proposed mechanism in scenario 3 has longer collaboration lifetime than the other two. And the effect of the proposed mechanism in scenario 1 is almost as same as the effect in scenario 2. Intuitively, more replaceable sensors and higher data aggregation compression ratio makes less communications traffic and more sleeping sensors. It is beneficial to save radio energy consumption. It extends the collaboration lifetime of the proposed mechanism.

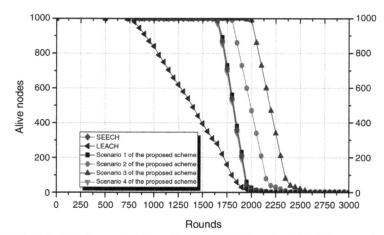

FIGURE 10.6 Lifetime evaluation of NLES, SEECH, and LEACH. In scenarios 1–3, NLES is more effective than the other two. In scenario 4, NLES is slightly less effective than SEECH.

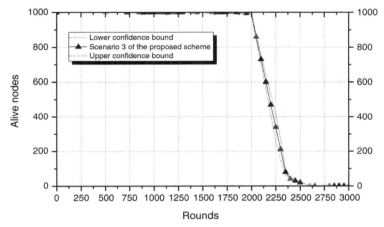

FIGURE 10.7 Confidence interval of the evaluation result in scenario 3. We set confidence coefficient as 0.95.

It is showed that the proposed mechanism has longer collaboration lifetime than LEACH. And in environment with lots of replaceable sensors and high data aggregation compression ratio (such as in scenario 3), the proposed mechanism has longer collaboration lifetime than SEECH. But in scenario 4, the proposed mechanism is slightly less efficient than SEECH. This is because the switches of replaceable sensors and data aggregation consume some energy. But because of low data aggregation compression ratio, only a little communication traffic is saved. This leads a little energy saved. The consumed energy is much more than saved energy in this mode.

From the result of scenario 4, the conclusion is drawn that NLES is not suitable for traditional CPS whose sensors and actuators cannot support VNFs. This is because NLES is relied on the advantages of NFV.

To estimate evaluation results for the true population proportion, confidence interval is introduced in this section. The confidence interval of the evaluation result in scenario 3 of NLES is illustrated in Figure 10.7. In this illustration, the confidence coefficient is set as 0.95.

10.5.3 Analysis of the Advantages for Traffic Volume Using SDN and NFV in CPS

Assume there are N sensors in context networks. With the central control feature of SDN, the proposed mechanism greatly decreases the message complexity. In the proposed mechanism, each sensor should broadcast a message to report their information. Therefore, the message complexity of the proposed mechanism is N. In Ref. [6] the authors declare the message complexity of LEACH is $N + K_{opt} \approx 1.1N$, where K_{opt} is the optimal number of cluster head sensors. And in Ref. [7] the authors declare the message complexity of SEECH is

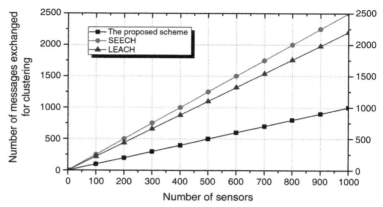

FIGURE 10.8 Message complexity evaluation of the proposed mechanism, SEECH and LEACH in mobile context. The proposed mechanism has lowest message complexity.

$N + 4K_{CH} + K_{CHC} + K_R \approx 1.25N$. But both the LEACH and SEECH assume the network is stationary. If they want to address the dynamic problem, the sensors must broadcast status information messages into the network. This doubles the message complexity of these two mechanisms at least.

The message complexities of SEECH, LEACH, and the proposed mechanism are illustrated in Figure 10.8. Obviously, NLES has exact lower message complexity than the other two. This greatly decreases traffic volume in the network than SEECH and LEACH, especially in large-scale CPS networks.

With the dynamic and instant deployment feature of NFV, more sensors have chances to go into sleep mode in CPS network. Also by this feature, CPS gets higher data aggregation and compression ratio for sensors, and thus decreases the entire traffic volume in the network.

10.6 CONCLUSIONS AND FUTURE WORK

This chapter proposed NLES to extend collaboration lifetime of CPS using SDN and NFV. The advantages of SDN and NFV techniques for improving energy efficiency for CPS network are analyzed. A new enhanced architecture that leverages the advantages of SDN and NFV is presented. In the enhanced architecture, topology of CPS network is controlled by logic controller that utilizes SDN technique and sensor functions are deployed dynamically into sensors in the network that utilizes NFV technique. By these means, topology control and sleep mode are introduced to extend the collaboration lifetime of CPS. As a result, the architecture provides more effective solution than traditional methods. Details about controller and sensors in the architecture are illustrated. Also workflows and protocols of mechanism are discussed. To obtain an effective topology decision, game theoretic approach is used to make topology decision that fully utilize the advantages of SDN and NFV and tackle the difficulties in complicated decision. The evaluation and discussion show that NLES has longer lifetime than existing methods. This is the first work using SDN and NFV to extend the lifetime for CPS, which is significant to improve the availability of CPS.

ACKNOWLEDGMENT

This work is supported by National Natural Science Foundation of China (Grant No. 61401273 and 61431008).

REFERENCES

1. Kang, W., Kapitanova, K., and Son, S. (2012) RDDS: a real-time data distribution service for cyber-physical systems. *IEEE Transactions on Industrial Informatics*, 8 (2), 393–405.

2. Li, H., Dimitrovski, A., Song, J., Han, Z., and Qian, L. (2014) Communication infrastructure design in cyber physical systems with applications in smart grids: a hybrid system framework. *IEEE Communications Surveys & Tutorials*, 16 (3), 1689–1708.

3. Ma, J., Lou, W., and Li, X. (2014) Contiguous link scheduling for data aggregation in wireless sensor networks. *IEEE Transactions on Parallel and Distributed Systems*, 25 (7), 1691–1701.

4. Zhu, C., Yang, T., Shu, L., Leung, V., Rodrigues, J., and Wang, L. (2014) Sleep scheduling for geographic routing in duty-cycled mobile sensor networks. *IEEE Transactions on Industrial Electronics*, 61 (11), 6346–6355.

5. Aziz, A., Ahmet, Y., Fitzpatrick, P., and Ivanovich, M. (2013) A survey on distributed topology control techniques for extending the lifetime of battery powered wireless sensor networks. *IEEE Communications Surveys & Tutorials*, 15 (1), 121–144.

6. Heinzelman, W., Chandrkasan, A., and Balakrisnan, H. (2002) An application-specific protocol architecture for wireless micro sensor networks. *IEEE Transactions on Wireless Communications*, 1 (12), 660–670.

7. Tarhani, M., Kavian, Y., and Siavoshi, S. (2014) SEECH: scalable energy efficient clustering hierarchy protocol in wireless sensor networks. *IEEE Sensors Journal*, 14 (11), 3944–3954.

8. Chiwewe, T. and Hancke, G. (2012) A distributed topology control technique for low interference and energy efficiency in wireless sensor networks. *IEEE Transactions on Industrial Informatics*, 8 (1), 11–19.

9. Anastasi, G., Conti, M., and Francesco, M. (2009) Extending the lifetime of wireless sensor networks through adaptive sleep. *IEEE Transactions on Industrial Informatics*, 5 (3), 351–365.

10. Kreutz, D., Ramos, F., Veríssimo, P., Rothenberg, C., Azodolmolky, S., and Uhlig, S. (2015) Software-defined networking: a comprehensive survey. *Proceedings of the IEEE*, 103 (1), 14–76.

11. Open Networking Foundation (2012) *Software-defined networking: the new norm for networks*, Open Networking Foundation.

12. Hu, R. and Qian, Y. (2014) An energy efficient and spectrum efficient wireless heterogeneous network framework for 5G systems. *IEEE Communications Magazine*, 52 (5), 94–101.

13. Zou, Y., Zhu, J., and Zhang, R. (2013) Exploiting network cooperation in green wireless communication. *IEEE Transactions on Communications*, 61 (3), 999–1010.

14. Han, B., Gopalakrishnan, V., Ji, L., and Lee, S. (2015) Network function virtualization: challenges and opportunities for innovations. *IEEE Communications Magazine*, 53 (2), 90–97.

15. Luo, T., Tan, H., and Quek, T. (2012) Sensor OpenFlow: enabling software-defined wireless sensor networks. *IEEE Communications Letters*, 16 (11), 1896–1899.

16. Matias, J., Garay, J., Toledo, N., Unzilla, J., and Jacob, E. (2015) Toward an SDN-enabled NFV architecture. *IEEE Communications Magazine*, 53 (4), 187–193.

17. Lin, Y., Lin, P., Yeh, C., Wang, Y., and Lai, Y. (2015) An extended SDN architecture for network function virtualization with a case study on intrusion prevention. *IEEE Network*, 29 (3), 48–53.

18. Abdulla, A., Fadlullah, Z., Nishiyama, H., Kato, N., Ono, F., and Miura, R. (2015) Toward fair maximization of energy efficiency in multiple UAS-aided networks: a game-theoretic methodology. *IEEE Transactions on Wireless Communications*, 14 (1), 305–316.

19. Guo, B., Guan, Q., Yu, F., Jiang, S., and Leung, V. (2014) Energy-efficient topology control with selective diversity in cooperative wireless ad hoc networks: a game-theoretic approach. *IEEE Transactions on Wireless Communications*, 13 (11), 6484–6494.

20. AlSkaif, T., Zapata, M.G., and Bellalta, B. (2015) Game theory for energy efficiency in wireless sensor networks: latest trends. *Journal of Network and Computer Applications*, 54 (1), 33–61.

21. Misra, S., Ojha, T., and Mondal, A. (2014) Game-theoretic topology control for opportunistic localization in sparse underwater sensor networks. *IEEE Transactions on Mobile Computing*, 14 (5), 990–1003.

22. Wood, T., Ramakrishnan, K., Hwang, J., Liu, G., and Zhang, W. (2015) Toward a software-based network: integrating software defined networking and network function virtualization. *IEEE Network Magazine*, 29 (3), 36–41.

CHAPTER 11

ENERGY AUDIT IN INFRASTRUCTURES

SHALIGRAM POKHAREL,[1] FARAYI MUSHARAVATI,[1] and
HOSSAM A. GABBAR[2]
[1]Department of Mechanical and Industrial Engineering, College of Engineering,
Qatar University, Doha, Qatar
[2]Faculty of Energy Systems and Nuclear Science, and Faculty of Engineering and
Applied Science, University of Ontario Institute of Technology, Oshawa, Canada

11.1 INTRODUCTION

As we have seen in previous chapters, energy conservation is an important aspect in building design. As technology advances, there are more opportunity to conserve energy either through supply management, for example, through the use of new power technology such as solar and wind, or through the replacement of existing energy end-use technologies such as using LED lamps for incandescent lamps or using smart air-conditioning systems instead of the current legacy air-conditioning system.

Energy supply management has been considered an important aspect in many countries through diversification of fuel and energy-generation technology. For example, due to the lack of its own resources, a relatively easy supply of fuel from its neighboring countries and through import from the third country, and the availability of new and efficient technologies, Singapore has moved its energy-generation system from diesel- and fuel-oil-based electricity generation to natural gas fuel generation since 2000. The use of natural gas in electricity generation has increased to more than 80% in a span of about 10 years. Supply management does provide an efficient way of generating energy. However, if its consumption is wasted due to the use of inefficient technology, then the attempt on supply management also gets wasted because supply technology alone cannot decrease energy consumption. This is a loss of economic and environmental opportunity.

Energy Conservation in Residential, Commercial, and Industrial Facilities, First Edition.
Edited by Hossam A. Gabbar.
© 2018 The Institute of Electrical and Electronics Engineers, Inc. Published 2018 by John Wiley & Sons, Inc.

Therefore, energy planners also look at demand management option, through the use of efficient technology to get maximum service with minimum energy input.

For energy supply, the primary energy is passed through several stages before it is finally utilized to produce end use energy. For example, electricity is used by the electric oven to provide heat in order to cook food in a household. However, the efficiency of energy utilization also depends on the type of utensils (devices) used and behavioral aspects of cooking. For example, a properly sized electrical rice cooker requires less energy to cook rice compared to cooking it in an electric cooking range. Therefore, it is necessary for us to know how much energy is being supplied to a building, what end use services are required in the building, and what technology and devices are used to convert (or distributed) energy from one form to another. New technologies may conserve energy but given the cost of resources (energy and technology), they may not be economically viable or may not provide enough incentives for building designers to switch to the new form of energy. Therefore, analysis of the economics of energy conservation is also equally important.

Given this background, energy audit can be defined as a process to identify the need to assess energy consumption in one or more energy end use services in a building or for a process, either on a regular basis or on requirement basis. Such a finding would help energy auditors to identify energy saving opportunities (ESO) or energy conservation opportunities (ECO) in the premise or the process that is being audited.

Energy audit usually requires an initial examination of energy consumption of an end use. That is, energy audit assesses the output of end use devices and compares the energy consumption with the benchmark or standards (if available).

Energy audits are conducted due to voluntary initiatives or mandatory requirements. For voluntary initiatives, there may or may not be a proper format or guideline to assess energy consumption. Also, there may be guidelines provided in each country to indicate environment-friendly energy consumption. Efforts such as LEED (leadership in energy and environmental design) certification manuals help in responsible energy use in buildings. These are proactive and voluntary measures to provide "green" methods for building design and construction, building operation and maintenance, neighborhood design, and small unit residential homes. Therefore, a voluntary energy audit can help identify processes that waste energy, and find better technologies (basically building operation and maintenance and small residential units) to achieve better energy conservation. There are various levels of certifications that can be achieved by the building owners based on the design and/or operation. In existing buildings, however, interior design and construction, and building operations and maintenance, for both large and small residential buildings, can be considered for the energy audit and resulting changes for "green" energy use.

Mandatory energy audits are imposed either by the national or local governments. They are usually periodic and can encompass the building design itself. For example, the government can regulate and audit water heating electrical equipment in a particular building. Mandatory audits are usually extensive and they are conducted with strictly outlined procedures. This type of audit helps the authorities to compare energy consumption per floor area or per process for the similar type of buildings and promote the development of better building design parameters, initiating "facilitation" programs to increase energy efficiency from the threshold levels or to ensure compliance with the standards set by the governments.

11.2 TYPES OF ENERGY AUDITS

In general, two types of energy audits are employed for voluntary assessment: walk-through (or macrolevel) assessment and detailed (or microlevel) assessment. Mandatory energy audits may also require these two types of auditing but when proper checklist and analysis processes are already established, the regular mandatory audit of buildings may only consider detailed audit of the energy system in the prescribed form and format.

Depending on the requirements, there could be several steps that are followed for both walk-through and detailed auditing. However, the primary activity in the walk-through audit is to start collecting gross level data, for example, the collection of monthly energy usage data and costs and other characteristics for an area of the facility in regular use.

As already mentioned, energy audit scope should be clearly defined. Audit for a building may be done only for one specific type of end use, for example, space conditioning or lighting. Therefore, the first thing to be done for the audit process is to establish the objectives of energy audit. The objectives should clearly outline the type and intensity of the audit and the expected outcomes from such an endeavor. Once the scope and objectives are identified, the next step would be to collect data: either gross data (for walk-through audit) or detailed data (for detailed audit). Although energy audits can be done by auditors, it is recommended to support the auditors by other professionals or persons such as facilities manager (for large building), or process engineer (for industrial processes), or household owners (for auditing of individual households). That means energy audits can be done in any kind of building or processes, for regular households, industrial buildings and industrial processes or commercial buildings, such as small offices to multistory shopping complexes or hotels and restaurants.

11.3 BUILDING DETAILS FOR ENERGY AUDITS

A commercial building has many energy end uses. Depending on the geographic location, the consumption for each end use can be different. For example, cooling load is higher in Middle Eastern countries such as Qatar, whereas heating load could be higher in countries such as Canada. End uses, such as lighting, although very important, can consume only a small portion of energy compared to other end uses such as electric motors, water boilers, space conditioners, and other devices consuming energy. A case example of lighting is considered here.

For a walk-through audit of lighting systems, basic information in terms of the number of lamps, the type of lamp, lighting area, energy used for lighting (if available), visual inspection of light color and intensity in a particular area would be required. There are standards available for use of light in different working conditions. A simple illuminator device can be used to measure lighting available at a particular location (or area of work). The illuminator can be used during the walk-though audit in order to find out the intensity of lighting in a particular place and could be the focus of attention for detailed lighting analysis. For example, if only a floor or area of the building is found to have lower illumination, detailed analysis for lighting can begin with that area.

11.4 BASICS FOR LIGHTING AUDITS

There are few guiding factors that should be understood while conducting the detailed audit analysis of lighting. These factors are in terms of type of lamps used (to assess luminous efficacy in lumens/watt, color temperature of lamp, and color rendering index), types of luminaires used, age of the lamp and luminaire, wall and luminaire surface, light maintenance schedules, area of the room, and height of the lamp from the working surface. Color temperature refers to the appearance of light generated by the lamp compared to that produced by the blackbody radiator at a particular temperature. For example, color temperature of daylight is about 6000°K. Color-rendering index (CRI), on the other hand, is the ability of the lamp to clearly distinguish different colors and are measured in terms of percentages. An incandescent lamp provides about 2800°K color temperature and has a CRI of almost 100%.

11.5 TYPES OF LAMPS

For generic lighting purposes, four major types of lamps are used: incandescent, fluorescent, compact fluorescent, and light-emitting diode (LED). A brief description of lamp types follows.

Incandescent lamps provide light when the filament inside the glass is heated in excess of 2500°C. These lamps provide 5–20 lm/W and have an operating life of about 500–2500 h depending upon its use. A continuous use of incandescent lamp degrades the filament faster and as a result, the operating life of the lamp is reduced. Although only 5% of the energy used by the lamp is converted to light, it is an excellent source of light with a good color temperature and has an excellent CRI, as mentioned earlier. Many countries have started voluntary or mandatory replacement of incandescent lamp due to their lower lumens output.

Fluorescent lamps (FLs) provide light due to the glow of phosphorus created by ultraviolet light generated by the inert gas and mercury inside a fluorescent tube. Electronic or magnetic ballasts integrated into the luminaire are used to start and regulate the fluorescent lamps. Ballast factor (BF) ranges between 0 and 1 and therefore, the lumen outputs are adjusted based on the ballast being used. For example, a 2000 lm/W with 0.9 BF produces only 1800 lm/W. For electronic ballast, BF could be more than 1. However, higher BF means more consumption of ballast energy and therefore, a good matching of lighting requirements with lamp and ballast becomes important. FLs can come either as straight tubes in U-shaped tubes. The color temperature of fluorescent lamp ranges from 2700 to 6500°K with a CRI of 50–98. Efficacy of fluorescent lamp can range between 30 and 100 lm/W. These lamps are measured in terms of the diameter of lamp tubes in terms of 1/8 in. That means the T12 tube is 12/8 in. in diameter. Operating life of a standard FL can range between 5000 and 20,000 h.

Compact fluorescent lamps (CFL) use the same principle for lighting as that for FLs. CFLs are available either in twin tube or double tubes and with or without integrated ballast. A CFL with integrated ballast is designed to replace an incandescent lamp in order to save energy used for lighting. As electric discharge in CFL is on one side only, it becomes easier to manufacturers to design CFL with different shapes. The color temperature of CFL is the same as that of fluorescent lamps. However, CRI for nonintegrated CFL is between 80 and 92, whereas that for integrated ones is between 76 and 82. Similarly, efficacy is also different for integrated and nonintegrated CFLs. Efficacy is usually lower for integrated CFLs. However, efficacy also depends on the color temperature of the light. The average operating life of the currently available CFL varies between 7000 and 20,000 h. However, in general, a value of 10,000 h is generally assumed as the operating life of a CFL.

LED is a semiconductor device that releases light from a chip. One of the constraints of LED is the heat generated by the semiconductor chip, which requires a sink and fast removal of the heat in order to prolong the operating life of a LED. Although LEDs are currently used in mass for small-scale applications, they are slowly replacing FLs and incandescent lamps as they can be packaged in different shapes. Most of the currently available LEDs have an efficacy between 20 and 90 lm/W. The color temperature of LEDs ranges from 3300 to 5000°K and CRI ranges from 70 to 90. In general, operating life of LED ranges between 25,000 and 50,000 h.

11.6 LUMINAIRES

Lamps are used in a particular type of lamp fixture, called luminaires as shown in Figures 11.1–11.4. Luminaires may be upward facing or downward facing, or may or may not have diffuser, reflector, or lenses. A good luminaire would reflect most of the lights generated by the lamp(s) in the fixture to the working surface, and the ratio of generated light to the light received in the working surface is called the coefficient of utilization (CU) of the luminaire. Therefore, depending on the lighting requirements and the type of reflectors used, more lumens might have to be

FIGURE 11.1 An example of downward light output in ceiling protruded luminaire.

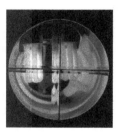

FIGURE 11.2 An example of downward light output with ceiling recessed luminaire.

FIGURE 11.3 An example of downward light output two T8 FL hanging luminaire.

FIGURE 11.4 An example of ceiling recessed luminaire with reflector (using four T8 FLs).

produced at the fixture in order to get the right amount of light on the work surface. However, luminaire not only reflects the light, it also reflects the heat generated in the fixture due to lamp, ballast, or lenses. However, when the lamps are used for prolonged hours, due to increased heat, the efficacy of the lamp itself can decrease. Therefore, designers should also consider heat removal if lights are to be used for long hours.

The four main factors that are considered for coefficient of utilization are the lamp lumen maintenance factors (to account for efficacy loss due to hours of operations and the use of ballast), luminaire maintenance factor (to account for dust accumulation on fixtures and lamps), lamp survival factor (to account for the number of lamps surviving after operating a certain number of hours), and the room surface maintenance factor. The loss of lumens is important because some studies have shown that there could be a gradual productivity loss of a person due to the poor lighting condition. Dust accumulation also depends on the type of activities conducted under the light, for example, lights in clean offices accumulate less dust compared to lights in the assembly line in a factory. A decreased lamp survival factor requires replacement of lamps frequently. Survival factor depends on the quality of manufacture and the operating hours of the lamp. In general, larger the operating hours, less would be the survival factor of lamps.

11.7 ROOM INDEX

Lumens required on the surface also depend on the height of the light from the working surface and area that it has to illuminate. Therefore, in some cases adjustments in height provide better illumination and fewer lumens are required for the same task. Standard tables may be available from the manufacturers of lamps and luminaires to obtain the light requirements for a particular room index. If L and W represent the length and width of the room and H is the gap between the working surface and luminaire, then room index (RI) can be obtained as $RI = [LW/(L + W)H]$. Room index is used for obtaining utilization factor of light in a particular room. The values of utilization factors (UF) are obtained from the manufacturers of luminaires or lamps. An audit can help to identify height that can provide maximum lumens required for a particular work activity.

11.8 EVALUATING THE NUMBER OF LAMPS REQUIRED FOR AN ACTIVITY

The evaluation requires measurement of illumination (E) to be received at a particular area (A) for a particular activity. Each lamp can be considered to produce ϕ lumens. Let the number of lamps per luminaire be n, and the number of luminaire be N. Then,

$$E = \frac{\phi \cdot n \cdot N}{A}$$

When we consider the maintenance and utilization factors, the equation needs to be modified as

$$E = \frac{\phi \cdot n \cdot N}{A} \cdot UF \cdot CU$$

It is clear from the above that more lumens output is needed at the luminaire in order to receive adequate illumination at the working surface.

The number of lamps required for lighting on each area can then be obtained as

$$n = \frac{E \cdot A}{\phi \cdot N \cdot \text{UF} \cdot \text{CU}}$$

It is assumed that the number of luminaires cannot be changed as they are fixed in the building.

11.9 ECONOMICS OF AUDIT IN LIGHTING

Energy auditing can be expensive for large-scale buildings, but it is worth understanding the energy savings opportunities and making the change. It should be noted that auditing may result in simple requirements such as regular maintenance of lamps and luminaire and replacement of aged lamps or luminaires or changes in the type of luminaire and type of lamps. Regulations or economics of such a change usually becomes the main driver for the decision on implementing the results of audit outputs. Audit may recommend removing some of the lamps due to overdesign of lumens for a particular work, or increasing the lumens with increasing number of lamps (which may require replacement of the luminaire), or wattage of lamp due to underdesign of lumens for particular work, adjustments in luminaire height, changes in the color of the room, replacement of old lamps with new efficient lamps (same shape and size), replacement of luminaires, lamps and ballast in favor of new and more efficient lighting technology, increasing daylight by making some minor structural changes, and finally, simply the maintenance of luminaires.

Various cost factors have to be considered for such a replacement. The cost of electricity, lamps, fixtures, and other relevant cost parameters should be obtained for economic calculation. Usually, management would like to see the difference in costs in the current state and the improved state. As financing for changes in lighting might have to be deviated from other more productive projects, a discount rate might be used to judge the net present value of the project. However, payback period calculation is also common to provide an indication as to the benefits of the new plan proposed through the energy audit.

It should be noted that prior to the calculation, details of lumens, type of lamps and luminaires, and other relevant information should all be obtained for retrofitting. Table 11.1 provides the features that can be used to evaluate the benefits of changes in lighting. It assumes that one type of luminaire is used for the whole building for simplicity purpose; however, calculations can be done with different types of the luminaire in different areas of the building.

TABLE 11.1   Parameters for Economics Calculations for an Office Building

For New design		For Existing Design	
i_n	New (n) type (i) of lamp	j_o	Existing (o) type (j) lamps
C_{in}	Unit cost of lamp	C_{jo}	Unit cost of lamp
C_{Nin}	Cost per luminaire	C_{Njo}	Cost per luminaire
ϕ_{in}	Lumen per watt (of lamp)	Φ_{jo}	Lumen per watt (of lamp)
W_{in}	Watt per lamp	W_{jo}	Watt per lamp
N_{in}	Number of luminaires	N_{jo}	Number of luminaires
L_{Nin}	Number of lamp per luminaire	L_{Njo}	Number of lamp per luminaire
W_{Nin}	Watts provided to each luminaire	W_{Njo}	Watts provided to each luminaire
O_{in}	Operating life of lamp	O_{jo}	Operating life of lamp
O_{Nin}	Operating life of luminaire	O_{Njo}	Operating life of luminaire
H_{Nin}	Light operating hours used per year	H_{Njo}	Light operating hours used per year
M_{Nin}	Materials required per luminaire (e.g., tools, wiring, tapes, and others)	M_{Njo}	Maintenance materials required per luminaire and lamp
F_{Nin}	Fitting cost of luminaire (labor cost)	F_{Njo}	Labor cost for maintenance
E		Unit cost of electricity	

Given the above parameters, the cost for new design can be evaluated in the following steps – a negative outcome of the evaluation would refer to savings from the new system:

1. Costs of new design – cost of existing design

$$\sum_i C_{in} \cdot L_{Nin} \cdot N_{in} - \sum_j C_{jo} \cdot L_{Njo} \cdot N_{jo}$$

2. Cost of new luminaire – cost of existing luminaire

$$\sum_i C_{Nin} \cdot N_{in} - \sum_j C_{Njo} \cdot N_{jo}$$

3. Cost of fitting and materials – cost of maintenance and materials

$$\sum_i F_{Nin} \cdot N_{in} + \sum_i M_{Nin} \cdot N_{in} - \sum_j F_{Njo} \cdot N_{jo} - \sum_j M_{Njo} \cdot N_{jo}$$

The cost of energy with new design − cost of energy with the old design. It is to be noted that if there is an energy demand cost charged to the premise, the reduction in wattage may also reduce the demand cost. However, in general, energy demand costs are charged on a flat rate basis and usually, they may not impact largely for lighting as the lighting energy required is a small proportion of total energy cost of a building. Therefore, energy demand cost savings is not mentioned here. This is to note that for a good comparison of saving, the same hours of lighting should be used. The division of 1000 is to account for unit charges, which are usually in kilowatt-hour basis.

$$E \cdot \left(\sum_i W_{Nin} \cdot H_{Nin} - \sum_i W_{Njo} \cdot H_{Njo} \right) \Big/ 1000$$

4. Lamps and luminaires (note that when the luminaire is changed, lamps are also assumed to be changed) are also to be replaced after the completion of their operating lives. Each set of lamp and luminaires have their operating lives, although the operating life of the lamp is much shorter than that of the luminaire. Therefore, once the luminaire is changed, the set of lamps are also changed and it would normally require a change in the lamps as well. For example, when LEDs are used, the fixture has to be changed once LED starts fading or degrading and similarly when LED fixture is damaged, the whole set of LED luminaire (along with LEDs) has to be changed. Therefore, for illustration purposes, only the changes in the lamps are considered here.

$$\sum_i C_{in} \cdot L_{iNn} \cdot N_{in} \cdot \frac{H_{Nin}}{O_{in}} - \sum_j C_{jo} \cdot L_{jNo} \cdot N_{jo} \cdot \frac{H_{Nin}}{O_{jo}}$$

Simple payback can be calculated by using the annual cost of implementing the system and the savings that may be obtained from the replacement of lamps and luminaire. It is to note that, in general, investments in lighting can payback within a year because very efficient lamps with high lumens per watt are available on the market. These lamps replace a number of old lamps (or watts per lamp) required to get the same level of illumination.

The analysis should focus on obtaining the total cost for the lifetime of the lamp. Economic analysis for comparison of two systems should assume the same life cycle and usually a constant cost. That means costs are pegged to the first year of analysis and lighting system with lower operating life should be repeated to match the life cycle of the lighting system with longer operating life.

ACKNOWLEDGMENT

This chapter was made possible by a NPRP award NPRP 5-209-2-071 from the Qatar National Research Fund (a member of Qatar Foundation). The statements made herein are solely the responsibility of the authors.

INDEX

Energy Conservation in Residential, Commercial, and Industrial Facilities, First Edition.
Edited by Hossam A. Gabbar.
© 2018 The Institute of Electrical and Electronics Engineers, Inc. Published 2018 by John Wiley & Sons, Inc.

IEEE PRESS SERIES ON SYSTEMS SCIENCE AND ENGINEERING

Editor:
MengChu Zhou, *New Jersey Institute of Technology and Tongji University*

Co-Editors:
Han-Xiong Li, *City University of Hong-Kong*
Margot Weijnen, *Delft University of Technology*

The focus of this series is to introduce the advances in theory and applications of systems science and engineering to industrial practitioners, researchers, and students. This series seeks to foster system-of-systems multidisciplinary theory and tools to satisfy the needs of the industrial and academic areas to model, analyze, design, optimize and operate increasingly complex man-made systems ranging from control systems, computer systems, discrete event systems, information systems, networked systems, production systems, robotic systems, service systems, and transportation systems to Internet, sensor networks, smart grid, social network, sustainable infrastructure, and systems biology.

1. *Reinforcement and Systemic Machine Learning for Decision Making*
   Parag Kulkarni

2. *Remote Sensing and Actuation Using Unmanned Vehicles*
   Haiyang Chao and YangQuan Chen

3. *Hybrid Control and Motion Planning of Dynamical Legged Locomotion*
   Nasser Sadati, Guy A. Dumont, Kaveh Akbari Hamed, and William A. Gruver

4. *Modern Machine Learning: Techniques and Their Applications in Cartoon Animation Research*
   Jun Yu and Dachen Tao

5. *Design of Business and Scientific Workflows: A Web Service-Oriented Approach*
   Wei Tan and MengChu Zhou

6. *Operator-based Nonlinear Control Systems: Design and Applications*
   Mingcong Deng

7. *System Design and Control Integration for Advanced Manufacturing*
   Han-Xiong Li and XinJiang Lu

8. *Sustainable Solid Waste Management: A Systems Engineering Approach*
   Ni-Bin Chang and Ana Pires

9. *Contemporary Issues in Systems Science and Engineering*
   Mengchu Zhou, Han-Xiong Li, and Margot Weijnen